FIRE AND BLOOD

FIRE AND BLOOD

Rubies in Myth, Magic, and History

Diane Morgan

Westport, Connecticut
London

Library of Congress Cataloging-in-Publication Data

Morgan, Diane, 1947-
    Fire and blood : rubies in myth, magic, and history / Diane Morgan.
        p. cm.
    Includes bibliographical references and index.
    ISBN 978-0-275-99304-7 (alk. paper)
1. Rubies. 2. Rubies—History. 3. Mythology, Hindu. 4. Mythology, Hindu—History. I. Title.
QE394.R8M67 2008
553.8'4—dc22   2007039086

British Library Cataloguing in Publication Data is available.

Library of Congress Catalog Card Number: 2007039086
ISBN: 978-0-275-99304-7

First published in 2008

Praeger Publishers, 88 Post Road West, Westport, CT 06881
An imprint of Greenwood Publishing Group, Inc.
www.praeger.com

Printed in the United States of America

The paper used in this book complies with the
Permanent Paper Standard issued by the National
Information Standards Organization (Z39.48-1984).

10 9 8 7 6 5 4 3 2 1

To Ariel Greenspoon, who is a shining and many-faceted gem in her own right.

Contents

Acknowledgments

Many thanks to my partner John Warner for his critical reading, helpful hints, and tactful editing of this book. And to the Praeger team, especially Elizabeth Potenza. Special thanks to Andrew De Young at Beacon for his patience and unfailing courtesy.

Introduction

Had we but world enough, and time,
This coyness, lady, were no crime.
We would sit down and think which way
To walk, and pass our long love's day;
Thou by the Indian Ganges' side
Shouldst rubies find...

"To His Coy Mistress," Andrew Marvell

The ruby is the world's most precious stone, and by far the rarest. Not even the diamond, with its unfathomable depths of light, nor the emerald, serene in its perfect greenness, is worth more, carat for carat, than the ruby. For the ancient Hindus, the ruby is Rajnapura: the Gem of Gems, at whose heart, it is said, surges an eternal and unquenchable flame. In Indian iconography it is the stone closest to the sun, the emblem of desire, the most sacred and holy of gems. It is no wonder that the ancient Hindus prized rubies well above all other jewels, calling it the king of gems (*ratnaraj*) or leader of precious stones (*ratnanayaka*). In English we call this most precious stone *ruby*, in Sanskrit it is *Manikya*, in Burmese *Budmiya*, in Cantonese *Se-fla-yu-syak*, in Arabic *Yakut bihar*. In all languages, priceless.

Yet for all its rarity, beauty, and passion, the ruby has always been a slightly sinister stone. It has none of the sublimity of the sapphire, the elegance of the emerald, the purity of the pearl, or the dazzle of the diamond. Of all the precious stones, the ruby is closest to a wild, living being: fiery, passionate, and dangerous. And while at times the flame within the stone seems to soar with the gay insouciance of a hearthside fire, more often it blazes with a passion that pulses, almost literally, in the veins of the earth. At times it burns, or even rages. The ruby is the shark of the gem world, which captures its prey by its ferocious beauty.

This is not a friendly stone of sky or earth or sea or river. This is a stone of fire and blood. Its color is traditionally like a drop of the reddest blood upon the neck of the whitest dove. And if it symbolizes blood, it has also demanded it. Its history is a trail of blood, tears, power, and passion. This is not a gem for sissies, but a stone of infinite power and magic. It is so beautiful and so deadly.

And for all its spectacular beauty, the ruby loves to hide—not just in the secret parts of the earth, but also in myth and history. From the fabled mines of Mogok in Burma to the Crown Jewels of England and Iran to the laser beam, the ruby has made an unforgettable journey, a journey of blood and beauty, magic and murder. Its end is not yet in sight.

1

The Conundrum of Corundum

Before there were any stories about rubies, there were rubies themselves. Simply put, rubies are the red gem variety of the mineral *corundum* (a word stemming from Tamil *kurundam*, the Hindi *kauruntaka*, and ultimately from the Sanskrit *kuruvinda*). The English word "ruby" simply means red stone, its name deriving from the Latin *rubeus, ruber,* or *rubrum,* all of which mean red. Sapphire is another variety of the same mineral.

Gem and mineral nomenclature is an odd thing. While "ruby" and "sapphire" are ancient and culturally honorable terms, they have no scientific basis at all. Scientifically speaking, there is no such thing as either ruby or sapphire. There is only corundum, which in its natural state is quite plain and colorless, and not used for gem-making. The stones we call ruby and sapphire are corundum with coloring added.

Technically, a "mineral" is considered to be a *natural, inorganic substance with a fixed chemical composition and a regular internal structure.* "Natural" is an important part of this definition. Thus, a synthetic ruby, whose chemical structure is exactly the same as a natural ruby, is in the rather odd position of being a "genuine" ruby but not a "genuine" mineral. It's a strange world we live in. Also, the definition makes it clear that a mineral cannot originate from an animal. Therefore, even though a pearl consists of a phase of calcium carbonate indistinguishable from the mineral aragonite, it is not "legally" a mineral, since it is the product of an oyster. Pearls, amber, jet, and coral are sometimes considered to be "gems," although usually that term is restricted to proper minerals. A few minerals, like gold or diamonds, occur as simple chemical elements, but the vast majority, like rubies, appear in compound form. It takes a precise combination of pressure and heat as well as the right mix of ingredients to make a mineral.

As mentioned, every mineral must have a definite chemical composition. (Rocks themselves can be made up of one or more minerals.) And while we say that the composition of a mineral is "fixed," it is allowable for some "substitutions" (actually impurities) to occur. These little additions can make or break a gem. When a drop of the right coloring agent is added, for example, both perceptions and market value change rapidly.

In the case of corundum, the magic formula is Al_2O_3, or aluminum oxide (two atoms of aluminum plus three atoms of oxygen) in crystalline form. It is one of the ironies of nature that two of the most common substances on earth combine (with a little dash of the right color) to form one of the earth's rarest and most precious stones. Aluminum oxide may sound like the siding rusting on a house, but don't be fooled. Such is the stuff of rubies and sapphires. It is true, however, that pure colorless corundum doesn't look like much (and doesn't fetch much commercially as gem material). In fact, the main use for colorless corundum is as a polishing compound and abrasive (it is very hard), such as emery paper, and for that purpose a large amount of it is mined every year. At one time, a major use for corundum (often inferior quality rubies) was as "jewels" for watches and other objects requiring very hard, wear-resistant bearings.

ALUMINUM, RUBIES, AND THE WASHINGTON MONUMENT

But aluminum oxide once had a cachet all its own, not because of corundum itself, but because of the aluminum that it contains. In fact, believe it or not, North Carolina corundum is closely connected with the construction of the Washington Monument. Atop the venerable structure rises a five-pound pyramid of solid aluminum—and the aluminum comes from rubies and sapphires mined in the United States.

When the monument was finished in 1885 (it was started in 1848), it was decided that it needed a special topper. After all, at the time it was built, the Washington Monument was the largest man-made structure in the world, and today remains the largest free standing masonry structure, weighing 90,854 tons and standing 555' 5 1/8" tall.

In 1885, aluminum, only recently discovered, had a certain mystique about it. It was tough, but lighter than steel. It was immensely resistant to corrosion. It conducted electricity, and so could be a good lightning-protector. It was beautifully lustrous and could be engraved. And at the time it was quite valuable—trading for about the same price as silver, about a dollar an ounce. What material could possibly more suitable? But there was a catch. Aluminum was hard to get, especially since the builders of the monument were looking for very pure metal. The most common source, bauxite, was not pure enough. (Modern technology, in the form of the electrolytic reduction process, can extract high-quality aluminum from bauxite, but this procedure had yet not been invented.) The only source for the high-grade aluminum needed was high quality corundum. And the nearest source for that was the Cowee River Valley in Macon County, North Carolina. It so happened that some of these gem deposits were operated by Tiffany for their use in jewelry and watches, so the famous jewelers entered the construction business.

The only way then known to extract aluminum from corundum was the complex and dangerous sodium reduction process, in which the crushed corundum was converted chemically into aluminum chloride and then reduced with metallic sodium to form salt and metallic aluminum. The main problem with this procedure (aside from cost, which was considerable) was that metallic sodium is a very nasty substance to work with, having the unfortunate propensity to burst into flame if exposed to air.

This was a job that obviously called for an expert, and in the case of the Washington Monument that expert was William Frishmuth (1830–1893), a Philadelphia metallurgist of great repute and a special secret agent to the War Department under Abraham Lincoln. (The foundry where he did his work is now preserved as a landmark in the history of metallurgy. It was the only aluminum foundry in the entire nation at the time.) In fact, it was Frishmuth who persuaded Colonel Thomas Lincoln Casey, who was in charge of the construction project, to use aluminum for the topper. Perhaps he felt that aluminum was a more progressive material than hoary old copper, brass, or bronze. In addition, its color would blend in handsomely with the gray granite of the monument. He promised he could make the crowning pyramid for $75.00, but in a classic early case of contractor overruns, ended up charging $256.10. The steamed Casey dispatched an assistant, Captain Davis, to go to Philadelphia to investigate the matter. Eventually, a price of $225 was settled upon. The final product was 22.6 centimeters in height and 13.9 centimeters at its base. It weighed 2.85 kilograms. Alas, Frishmuth came to a rather bad end, dying of a self-inflected gunshot wound, but he is not forgotten. Every year the American Foundry Society presents the William Frishmuth Award to the "Foundryman of the Year."

In 1884, before the pyramid was actually placed on the monument, however, Tiffany borrowed it and put it on the floor of the Fifth Avenue store as a publicity gimmick. (People could step over the pyramid and then claim that they had "stepped over the top of the Washington Monument.") The capping ceremony on December 6, 1884, introduced the strange metal to millions of people who had never heard of it, and they appeared suitably impressed with the "new" metal. (The monument was not formally dedicated until the following year.)

The whole monument was refurbished between 1996 and 2000, with, in a fitting nod to American capitalism, Target footing part of the bill. The project was not named "From Tiffany to Target," although perhaps it should have been.

FROM MINERAL TO GEM

Being a mineral is not such a matchless achievement, despite the attention lavished on a few species. There are over 4,000 different named minerals, with between 50 and 100 more discovered every year. Some minerals are so scarce that there is only one known specimen; these minerals are more curiosities than stones with dependable commercial value. These special stones are snapped up by collectors and museums for display, along with other rare minerals. True collectors don't make a hard and fast distinction between commercial, subjective terms like "valuable" and "not valuable." They are more interested in words like "rare," "unusual," and "interesting." Therefore, most collectors wouldn't bother displaying an ordinary cut diamond that might be "worth" thousands of dollars on the gem market. They are too common and not mineralogically interesting. A more common stone embedded in a rare or peculiar matrix would excite much more interest among collectors.

Of the 4,000 recognized minerals, only about 50 are considered "gems." There is no objective standard for what distinguishes a gem from other minerals, but it is commonly agreed that gems are minerals which are beautiful, durable, and rare. Of these gems, only about 15 are in common commercial use. Some truly spectacular gems are not used as jewelry because they are so *extraordinarily* rare (such as bixbite or "red emerald") that there is no real commercial market for them. Rubies, though, are gems in every sense of the word: they are tough, gorgeous, and rare enough to be coveted, but still easily available to interested buyers. The durability factor is rather important, too, as a stone needs a certain amount of toughness to be worn as jewelry, especially as a ring. Beautiful stones that are too soft to be worn as jewelry, like very rare stones, usually end up in the collector's case.

Gems are also portable, a feature that throughout the centuries has made them a handy source of wealth for people on the run. From persecuted Jews to disgraced French nobility to panicky Russian émigrés, gems have been the means to begin a new life in a new land. They are the universal currency.

The fact that rubies and sapphires are merely varieties of the same mineral was not discovered until the eighteenth century. Sapphires and rubies, as befits close relatives, are often found in close proximity, sometimes even in the same mine, particularly in Burma and Sri Lanka. In other places, such as Madagascar, Kenya, Thailand, and Tanzania, both gems occur, but at considerable distances from each other. Still other places, such as Australia, China, and Montana, produce sapphires only. Large sapphires are much more common than big rubies, as they develop from the mineral "melt" earlier than rubies. Since sapphires are much more common than rubies of similar size, they trade at a much lower price.

Because the word "corundum" strikes no fire at the heart, the red variety of this mineral has always been called ruby, while the other varieties are known as sapphire: pink sapphire, yellow sapphire, and so forth. Those sapphires with colors other than blue usually are called "fancy sapphires." This seems wrong, at least to me, since sapphire registers "blue" in the minds of most people. By all rights, pink corundum should be called just that—not pink sapphire. However, "corundum" is a word with no cultural resonance. The term "pink sapphire" apparently has a lot more cash value.

Curiously, what created the color (and hence the value) of the stone, are paradoxically, in the purest sense of the word, *impurities*, resulting from stray atoms of other elements getting mixed into the corundum.

As mentioned earlier, plain unadulterated corundum is colorless. For corundum to turn red and thus earn the name of ruby, the magical coloring agent is chromium (or more rarely, vanadium). Chromium is a strange substance. It makes emeralds green, too. It really doesn't take very much chromium to do the trick. The precise amount needed to color a ruby red is disputed, but it is very, very small, even for the reddest stone. One educated guess is between 0.10 and 1.6 percent of the crystal lattice. The amount of chromium determines the color saturation. Just a little bit of chromium will produce the so-called pink sapphire, which is really nothing more than an insufficiently colored ruby. A little more will result in a fiery ruby. The "best" or pigeon blood colored stones contain about 0.30 percent pure chromium oxide. If the oxidation condition changes, the stone will take on an orange shade.

Ruby's more modest sister, the blue sapphire, receives its color from the simultaneous addition of iron and titanium. If you add iron alone to corundum, you will have a "yellow sapphire." Sapphire of a more golden hue has iron and chromium. Other additives will turn the stone orange, yellow, green, gray, or even multicolored.

Sometimes a few iron atoms also sneak their way into rubies, and that can be a bad thing for the red stone. If too much iron gets into a ruby, it develops a rather dingy or brown appearance—one reason why Thai rubies tend to be inferior to those from Burma, which are pretty much iron free. Vanadium will edge the gem to the blue side, an effect some people like and some don't. Those who like it claim that a true "pigeon blood ruby" has the slightest of blue secondary effects. Those who don't like a blue undertone demand a pure, unadulterated red.

The four great gemstones, which combine the salient features of beauty, rarity, and durability, are the diamond, the emerald, the sapphire, and the ruby. Of these four major gems, the ruby is the rarest, and by a good margin. (That may not have always been true. According to one myth, rubies once littered the ground of the Garden of Eden, making an uncomfortable surface for barefoot walking. Garden of Eden rubies, if their provenance could be proven, would certainly be highly collectible today as historical curiosities, regardless of the objective quality of each gem.)

It was once impossible to determine with certainty where in the world a particular ruby was born. However, this can be done today, although it's still not particularly easy. The key is usually to carefully examine the trace element content of a gemstone and compare it to that of stones of known provenance. Even gems that form in identical parent rocks, for example in dolomitic marbles, still show specific local differences that can tell the trained observer if the stones came from Jagdalek in Afghanistan, Mogok or Mong Hsu in Burma, Chumar or Ruyil in Nepal, the Hunza Valley in Pakistan (whose rubies are as fine as Burmese ones, but much less common), Morogoro in Tanzania, or Luc Yen in Vietnam. (While rubies have been sold to finance drugs and wars, they have largely escaped the opprobrium awarded to so-called "conflict diamonds," because unlike diamonds, the country of origin is traceable, although laboratory testing may be necessary.)

At one time it was impossible to tell the origin of any particular stone, but that has changed with the development of "DNA fingerprinting" for rubies, sapphires, and emeralds. It turns out that the water in which the gems crystallized so many millions of years ago varied widely in its mineral content. The DNA process can take a small sample of the stone, vaporize it, and measure the oxygen isotope ratio.

There are other ways as well, which can be done without destroying the stone. Subtle, noninvasive techniques, such as particle-induced X-ray emission (PIXE), may be used to examine all elements heavier than sodium. The identification of inclusions can be complementarily performed by Raman micro-spectrometry. The great importance of this is not for the gem trade, however, but for archaeology, where the technique has been used to study rubies from a Mesopotamian statuette and other artifacts, helping scientists and historians understand early trade patterns. Still, it is also important to buyers of important stones. A good Mogok ruby will sell for more than an equally fine gem from somewhere else—it's the romance of the name. (Gems from Kyatpyin, Sagyin, Nysaseik, Kathe and other areas strictly outside the Mogok district are still considered Mogok gems, as they share specific geological traits, although most of them are not of true Mogok quality.) However, Mong Hsu rubies are not properly Mogok gems, although they are, of course, Burmese. Gem laboratories that currently perform country-of-origin tests in the United States include the American Gemological Laboratory and the Gemological Institute of America's Gem Trade Laboratory. (Established in 1931, the GIA is the world's largest nonprofit institute of gemological research and scholarship.) For the average gem buyer, however, the source of the stone is unimportant, just as most people haven't the "eye" to discern subtle differences in color or tone. The ruby is a culturally critical gem, whose history has collided with ours—and made the fire-red sparks fly.

2

Birth of the Stone

At this very minute, somewhere, new rubies are being formed, although not quickly enough, of course, to satisfy their potential owners. However, the birth of the stone is not often a smooth uninterrupted flow. Gem growth occurs in a desultory fashion, with stops and starts. In some cases, the gem may actually "deteriorate" a little before growth begins again. In some cases, foreign material gets trapped in the growing gem, creating what jewelers call a "flaw" in the mineral.

Ancient Hindu myth informs us that the ruby burns with an inextinguishable internal fire, and perhaps that concept is not so far adrift from the scientific truth of its rare birth. The ruby is indeed born in fire, although the details of its conception and parturition are still mysterious.

One of the most revered Indian sources for ruby lore is the semi-sacred Hindu epic *Garuda Purana* (c. 400 CE). This particular purana is written in the form of instructions by the benevolent god Vishnu to his sacred swan, also named Garuda. The work deals with astronomy, medicine, and grammar, as well as with the structure and qualities of gemstones. This epic is also noted for its various depictions of hell. In some parts of India, the book is read over the bodies of the cremated dead, and it has some fascinating images of exactly what will happen to sinners when they die. Still, despite its many morbid passages, the book is considered a lucky gift, if accompanied by a golden image of a swan.

The *Garuda* warns against flawed rubies of any kind. A passage in Chapter 70, supposedly the advice of the great sage Sri Suta Goswami, asserts:

A ruby, although genuine, should not be worn if it has strong color banding, excessive inclusions within like numerous internal cracks, a sandy appearance, a rough surface, or is dull and lusterless. Anyone using such a flawed ruby, even out of ignorance, will suffer from disease, or loss of fortune.

A flawed ruby will also spell danger for one's relatives. Even worse, the *Garuda* states that "a spurious alien gem can nullify the potency of many gems endowed with good qualities." Further elaboration on this concept continued down the ages. Owning a dull ruby would guarantee problems with one's siblings, and a brittle one might mitigate against having sons, while a cracked stone brought general bad luck. A stone with bubble inclusions was rendered ineffective, while a dusty ruby augurs stomach problems and infertility. Most serious of all, perhaps, was a "flimsy" stone, which guaranteed the wearer would be struck by lightning!

On the other hand, the wearer of an unflawed ruby would get rich, have a multitude of fine children, and achieve happiness, honor, and respect. It is not clear as to whether you can borrow a ruby to attain these worthy goals, or if you actually have to own a flawless ruby (which suggests that the wearer is already rich).

The mysterious ruby indeed becomes more richly hued in certain conditions. An intense light produces an intense color, while ordinary light reveals a more mundane looking stone. (The ancient Hindus knew their rubies. Although gem quality ruby is not found in India, entire communities of merchants such as the Chettiars of the southern Indian state of Tamilnadu made their living and their fortunes by trading in the Burmese ruby.) Owning a good quality ruby, according to Indian legend, will ensure your rebirth as an emperor, although a mediocre one may only get you reborn to a lower kingship. The current comparative lack of royalty of any stripe may indicate the falling quality of rubies worldwide.

The *Garuda* calls rubies

...beautiful and effulgent...possessing manifold virtues....The rubies from these fragrant lands are found in a variety of hues....Some are like human blood.... Being illuminated by rays of the Sun, this crystal species shines forth with wonderful color and brilliance.

According to the *Garuda*, rubies were born in the following way. The mighty demon Vala set about wreaking havoc in the heavens, mostly by stealing cows, a serious offense among Hindus. He didn't steal regular Holsteins or even Brahmins, though, but special "cloud cows," which if anything are even more holy than the earthly ones. The angry gods destroyed and dismembered Vala, but demons cannot be completely annihilated. They always turn into something else. His body became the seeds of all the gems we know today. In some versions of the tale, the parts of the dismembered body were scattered when all the creatures of the earth rushed up to gather their share of gem seeds, with the blood-rubies getting carried off to India, Burma, Afghanistan, Pakistan, Nepal, Tibet, Sri Lanka, and Thailand.

In another version, the Sun God Surya let drops of Vala's blood fall into the deep pools of the mythic Bharata (an area comprising modern Burma, Siam, India, and Sri Lanka), where they turned into rubies. A rival myth suggests that rubies were the drops of Surya's own blood. Surya is mixed up in a lot of ruby myths—I tell another one a little later. In any case, the point is made and reiterated both in Jyotish (the Hindu system of astrology) and in Hindu Ayurvedic medicine that there is a direct, vibrant link between this stone and the sun god, who has passed on his many powers and strengths to the ruby.

The teeth of Vala ended up as pearls in the oceans of Sri Lanka, Bengal, Persia, and Indonesia, and like all pearls have a strong connection to the moon. "The teeth of the demon Vala...fell like stars into the oceans below and became seeds for a species of gems with the luster of beams from Chandra, the Full Moon. Entering into the shells of oysters these seeds became pearls." (According to this Purana, by the way, there are other sorts of pearls besides oyster pearls—there is the Conch Pearl, Cobra Pearl, Boar Pearl, Elephant Pearl, Bamboo Pearl, Whale Pearl, Fish Pearl, and Cloud Pearl, nine kinds altogether.) These pearls are unknown to modern science.

Vala's skin turned to yellow sapphires and ornamented the Himalayas. His ruddy fingernails changed to hessonite garnet seeds that fell into the lotus ponds of Sri Lanka, India, and Burma. His toenails became the seeds of red pyrope garnet. His bones changed into diamond seeds. His eyes transmogrified into shining blue sapphires.

Even the demon's internal organs were put to good use. His bile became emerald seeds, his intestines coral. The fat from his body turned to jade; his semen became quartz. The rosiness of his complexion became "bloodstone coral." And even his cries turned into the seeds of the cat's eye chrysoberyl, an idea possibly suggested by the yowl of a cat in season. Vala is also conveniently, the "soul of the cave" in Indian mythology (the very name means "cave" and is the source of our word "valley"), and caves are the often the origin of gems. They are also holy places, as evidenced by various cave temples such as the ones in Ellora in western India, and made famous in Forster's *Passage to India*.

The Hindu myth doesn't say anything about rubies ending up in Europe or South America, and this much is accurate. Rubies did not appear in Europe until Graeco-Roman times, and they were imported from the East, not mined locally. And of course, the Indians had never heard of the Americas.

ST. HILDEGARD

Another proto-scientific explanation of the birth of crystals, this one European, is provided in writings attributed to the medieval mystic Hildegard of Bingen (1098–1179), who wrote: "Gems have their origin in the East, and in especially torrid zones. There the sun heats the mountains like fire, and the rivers are always boiling hot...Where the waters touch the burning-hot mountains, they foam." The foam then solidifies and turns to crystal. The temperature at which the foam dries determines both their color and the powers of the resultant gems. "Thus gems originate from fire and water, and for this reason they also contain heat, moisture, and many powers." These ideas are not so far removed from modern scientific thought as it may first appear, as indeed gems are born in heat and borne by water.

A few scholars are not convinced the scientific works attributed by Hildegard really belong to her, after all, as their style is completely different from the ecstatic, visionary writing for which she is better known. (She suffered from acute migraine headaches.) However, it is perfectly reasonable to suppose she could accommodate her writing style to agree with her subject.

One of her most notable contributions to science writing was her description of the female orgasm, the first in Western literature:

When a woman is making love with a man, a sense of heat in her brain, which brings with it sensual delight, communicates the taste of that delight during the act and summons forth the emission of the man's seed. And when the seed has fallen into its place, that vehement heat descending from her brain draws the seed to itself and holds it, and soon the woman's sexual organs contract, and all the parts that are ready to open up during the time of menstruation now close, in the same way as a strong man can hold something enclosed in his fist.

She believed that the "strength of the semen" determined the baby's sex of the child (strong semen equals male children, of course), while the amount of love and passion involved were responsible for its temperament. The only connection all this has with rubies is that rubies are stones of passion, and the description is too good to leave out.

RUBIES AND ROCKS

While ancient Indian lore maintains that God created the ruby and then made mankind to enjoy it, geologists prefer a more down-to-earth explanation. Making such an enduring marvel as a ruby is not easy, even for wise and clever Mother Nature. Like most gemstones, rubies are found in rocks. While there are three main "families" of rock, only igneous and metamorphic rock play a part in the formation of rubies. (A third member of the ancient and estimable rock family, sedimentary rock, is not productive of gemstones other than opals. Although occasionally gems are found in sediment, these would be a "secondary deposit," meaning they were not actually formed there.)

Igneous rocks, whose name means "fire," are the world's oldest rocks, and are indeed born from volcanoes and magma, the molten material that surges beneath the surface of the earth. (Magma contains more than liquid, though; it also contains gas and minerals.) Magma is usually less dense than the surrounding rock and tends to bubble to the surface. When the magma pours through openings in the earth, it is called lava, and the rocks that form from it are known as "volcanic" or "extrusive" igneous rocks, as they are *pushed out* from the earth. The magma, which can form into conical basalt shields (basalt is the most abundant igneous rock), contains many minerals that were actually formed deep inside the earth, including sapphires and rubies. (It is not actually possible to observe this process, of course; people are simply extrapolating from available evidence.)

About 90 percent of the earth's crust is made up of igneous rock, and igneous rock constitutes everything (in the form of melts) that lies below the crust. Those rocks that solidify below the surface are called "intrusive" rather than "extrusive" rocks; they are also known as *plutonic* rocks, named after the Greek god of the underworld, Pluto.

As the magma cools, the crystals begin to form—and the more slowly the magma cools, the bigger the crystals tend to get. That's because as the magma cools, it thickens, making it increasingly difficult for the chemical constituents of the minerals to flow through the magma to places where the crystal growth is taking place. The longer the magma stays hot and liquid, the more time the ingredients will have to reach the growth spot and the bigger the crystals will be. Large crystals also need "room to grow," and are thus most likely to appear in fissures or cavities of rocks.

Many rubies are found in the kind of igneous rock known as pegmatite. Pegmatite is a once-fluid, coarse-grained granite comprised of a variable mixture of mineral aggregates. Pegmatite is widely distributed in the earth's crust and has been observed in chunks as big as 3,000 meters long and 700 meters wide. Corundum often develops in aluminum-rich, silica-poor pegmatite and is carried to the earth's surface by basalt during tectonic plate movement. (Silica is a compound formed from silicon and oxygen. It is the basis of many minerals, including quartz and the gems opal and amethyst.)

The oldest pegmatite deposits, located in Canada, Greenland, and Russia are nearly three billion years old, almost as old as the earth's crust itself, while the youngest deposits (Himalayan), where rubies are often found, are a mere five million or so. Yet rubies do not age. The fire that was locked in their hearts millions of years ago still burns, even after the emperors and empires that fought for them have crumbled to dust and ashes.

Both Burma and Sri Lanka are quite famous for their gem gravels, which contain not only rubies but also sapphires, spinels, garnets, chrysoberyls, topazes, and tourmaline. Since most of the stones have been tumbled about in water and bounced against other rocks, many are somewhat the worse for wear, and so often only the toughest, highest quality stones survive the trip. The poorer stones have already been discarded by the forces of nature. In fact, some geologists consider the stones found in these deposits to be a product of a kind of "natural selection." Small and flawed stones generally don't make it through the entire process. Alluvial stones are also much easier to retrieve than stones locked in the bedrock, so they are indeed a gift from nature.

When rubies are born, they take shape as hexagonal crystals with flat or tapering ends. Some gems are found right at the surface of the rock—others of the same variety are buried deep within it. Still others are weathered free from the rock that locked them in, washed down streams and rivers draining from volcanoes, and deposited in gravel, sand, or clay. As rubies are heavier than most other minerals, they don't travel far and may be deposited in large numbers (gem gravels) close to their source. In some cases the depositing rivers no longer exist, but the stones remain where they were dropped. More rubies are found this way than any other. How much gets deposited and where depends upon the stream. The faster the stream and the more turbulent its flow, the more stones it can pick up. When the stream loses velocity because of a change in gradient or other reason, its "transporting power" is reduced and it will drop part of its load.

Secondary deposits such as this are called alluvial, or placer, deposits and are a main source of many heavy minerals like gold, diamonds, platinum, and rubies; in fact, some of the world's greatest stones were discovered in dried up ancient riverbeds. The word "alluvial" means carried by water. Alluvial stones are washed out from their ore by streams, and these are the ones alluded to in Marvell's poem at the beginning of this work. Alluvial gravels, historically, have been the most common source for gem material.

These are "secondary deposits"; in other words, the gems are not originally from the riverbed or gravel deposit but from a source upstream. They are washed down to their present location as a consequence of erosion or weathering, which is defined as the breakdown of rock exposed to air, water, or certain biological activity, like a tree root breaking through rock. The original homes of rubies are bands of crystalline limestone, often along with feldspar, garnet, graphite, mica, spinel (with which rubies were historically confused), pyrrhotite, and wollastonite.

One Burmese tale relates how a royal grandson of the legendary eleventh century King Anawrahta (1044–1077) was out hunting aboard his trusty elephant. During the hunt his party spied a mass of alluvial gems so large that a boat six cubits long and four cubits broad was required to carry the stones back to the Palace. (A cubit is about as long as the distance from your elbow to the end of your middle finger. Tape measures were once hard to come by.) The largest rubies were placed in the relic chambers of two pagodas built in honor of the find. Not many people today find rubies quite in this quantity, in fact, no one does, but the story captures the dreams of river panners everywhere.

When ruby-bearing rock is discovered intact, the stones can be removed by drills or picks. In other instances, large scale mining operations are commercially viable. These operations remove large quantities of rock, then crush, wash, and pick through it to find the gems. However, this method is so expensive and time-consuming that it usually makes more economic sense simply to collect rubies from gem gravels in time-honored fashion.

Igneous rocks are only one source for rubies, however. They can also form in aluminum-rich metamorphic rock. Metamorphic rocks are so designated partly on the basis of their texture, and partly on the basis of the minerals that comprise them. Most metamorphic rocks derive from shale, sandstone, or limestone. The word "metamorphic" comes from the Greek words *meta*, meaning "after," and *morphe*, meaning "shape" or "form." So, metamorphic rocks have been changed in shape or form. Metamorphic rock derives from igneous or sedimentary rock (or even another kind of metamorphic rock) that has been subject to great pressures that change its character. In the case of rubies, what has often occurred is the heat and pressure caused by the squeezing and buckling that went on during the formation of mountain ranges, causing the horizontal sedimentary strata to evolve into metamorphic rock.

As the rock is metamorphosing, gems like ruby and sapphire and garnet can form within them. Ancient metamorphic marble (more than 500 million years old), which forms from limestone that has been subjected to great pressure, may contain rubies of very high quality, as in the case of Burmese rubies. Marble is derived from limestone, itself made from the shells of countless sea creatures deposited on the ocean floor over millions of years. Heat and pressure changes limestone into the metamorphic rock marble. Besides metamorphic marbles, ruby sources include gneiss (a banded, granite-like rock) and amphibolite.

How did this real metamorphosis occur? In the case of rubies, one great metamorphic pressure developed when 50 to 60 million years ago, the then-island continent of India slowly slid into the rest of Asia, pushing up the mountains we call the Himalayas. Tibet was lifted 16,000 feet by the process. This collision provided the heat and pressure needed to change limestone into marble and force magma into the marble to make rubies.

Ruby deposits in some areas are present as inclusions in marble formations. (Marble is in itself a valuable resource, and is quarried as a building material. It makes marble even more interesting that occasionally rubies turn up in it.) These marbles are very poor in iron, which is a good thing for making rubies, since when iron gets into the corundum crystals, it dilutes the red, and imparts a brownish coloration, as is the case with many rubies from Thailand. Like pegamatite, marble has the advantage, of not only being low in iron, but also low in silica, one of the most abundant substances in the earth's crust; however, corundum does not form properly in the presence of large amounts of silica. Rubies are rare also partly because their coloring agent, chromium, is not normally a component of the rocks involved in the crystallization process.

DEMONS, CROCODILES, AND MAGIC SNAKES

Like the ancient Indians, the Burmese had some interesting concepts about ruby birth that had nothing to do with pegmatite intrusions.

According to Burmese legend, there once existed a gigantic serpent-dragon, Naga, who laid three eggs. From the first was hatched Pyusawti, the King of Pagan, a city in the middle of Burma founded in 849 c.e. and which was destined to be the capital of the country for 250 years. Before Kublai Khan took the place over in 1287 it was said to contain nearly 13,000 pagodas. Remains of about 5,000 of them can still be seen. Modern Pagan, however, is little more than a village.

From the second egg emerged the Emperor of China, and from the third came a wondrous seed from which in turn sprouted all the rubies in the world. I should note that while everyone agrees about the third, gem-bearing egg, there is some debate about the contents of first and second eggs.

Various other proposed hatchlings include the future prince Sawhti, a beautiful maiden, a fierce tiger, and a giant crocodile.

But the crocodile story cross-fertilizes that of the ruby story, just to show how complex these tales of origin can get. One crocodile folk tale is particularly instructive. It is the story of Ngamoyeit, whose egg was found by some kind fisherfolk and placed in a nearby pond. Imagine their surprise when the egg hatched into a crocodile. (It is not clear what they might have been expecting. Crocodiles are common in Burma, after all, and do hatch from eggs. Besides, the egg was too big to belong to a robin, for instance.) The undaunted couple fed the crocodile every day until nature took its course, and the crocodile killed and ate the husband. Of course, the tale does not end there, not in a culture where everyone is reborn again and again. The fisherman was reincarnated as a powerful magician determined to wreak vengeance upon the thankless beast. By this time, however, a hundred years had passed, and true to Burmese legend, the crocodile had achieved humanhood. (All crocodiles turn into human beings when they reach the century mark.) Still, the former croc was doomed. The fisherman-turned-magician struck the water three times with his magic wand. "Oh, rats," thought Ngamoyeit, "Now I have to return to the fisherman to be killed." And so it was. As he died, however, the part of the crocodile's body that had been in the water turned to rubies; the part on the riverbank turned into gold. Presumably the creature had lost his human form and returned to crocodile status at the moment of his death.

A similar egg-and-Naga legend tells us that long ago, some mischief-making spirits carried away a hunter. When they reached the place where the Naga had laid her precious golden egg (only one egg in this story), the hunter snatched it up and took it away. Somehow he also seemed to have gotten away from the kidnapping spirits. But while he was crossing a stream in "Mogok Kyappyin," a sudden current knocked him over and he dropped the egg. It broke and became a wealth of iron and rubies. This tale accounts not only for the streambed location of many rubies but also indicates how valuable iron was to ancient people once they learned how to smelt it.

Both tales show us that the ancient Burmese, along with many other peoples, believed that gemstones were more akin to living organisms than to rock, and were "born" rather than "made." The ruby's color, for instance, was due to its gradual "ripening," and the redder it became, the riper it was. (The ancient Inca had similar ideas about emeralds.) If you found a pink stone, you could replant it in the earth and wait until it ripened some more, although it is not stated as to how long this ripening might take. If the stone became overripe, on the other hand, it developed cracks and fissures.

The idea that rubies were equivalent to some sort of fruit got a full treatment in the mythical account of the Kalpa Tree, the wish-fulfilling spiritual tree, which grows on an island in the Mandara Forest. This fabulous tree (which is often depicted as growing upside down) is inhabited by the god Shiva and is made of precious stones—pearls, emeralds, and coral. Its fruit, of course, consist of rubies.

Another ancient Burmese fable claims that the rubies of fabled Mogok were actually first discovered by an eagle, many, many years ago, before the birth of the Buddha. According to this story, the largest, strongest, and oldest eagle in all creation was zooming through the air when he spied (with his eagle eye, of course) what looked like a luscious, bloody fresh morsel of meat upon the hillside. The eagle swooped down and snatched it up. But lo and behold, even his powerful talons could make no impression upon the obdurate stone. Again and again he tried. Finally, the eagle realized that this was not London broil but a matchless stone. The eagle then carried it off to his eyrie high on the highest mountain, where undoubtedly it still remains. (Precisely the same story is told about diamonds. Marco Polo [ca. 1254–1324] records the entire tale in his writings. The only difference is that in that case diamond-finding eagles are white.)

Here's a similar tale from Burmese legend, one which combines birds and Nagas. A female Naga named Zathi came to the world to bring virtue. She fell in love with the Sun Prince and begat an egg by him. The Sun Prince departed before the child's hatching, but wrapped a ruby (always the symbolic stone of the sun) in a silk bag and gave it to a *white crow* to pass along to Zathi, to buy a kingdom for the future child to rule over. The crow became distracted by hunger, however, and left the ruby-bag in the fork of a tree where it was later discovered by a hunter, who vowed to find the source of the stone and went looking for it. Whether he was successful or not is uncertain. An alternate story says the thief was a merchant who replaced the ruby with some dung, which the unsuspecting crow presented to the Naga. She was not at all pleased and slid into the lake, abandoning her precious egg. The mountain spirits took pity on the motherless egg and washed it into the Irrawaddy, where it opened and spilled out millions of gems. The Sun God was so angry about the crow's dereliction of duty that he cursed it and turned it black.

THE CRASHING CONTINENTS

The eastern part of Burma, where the best gems are, lies on the Southeast Asia tectonic plate, while the western part really belongs to the Indian plate.

The rocks of the western half of Burma were deposited during the Tertiary and Cretaceous periods (about 65 million years ago), but those in the east (the Shan Plateau and Tenasserim) are even older.

Before the acceptance of plate tectonics as a mechanism for continent formation, scientists believed that the earth's interior was cooling off and contracting, which forced parts of the surface to buckle and make mountains. While this theory explained certain features of the earth's shape, it didn't explain rift valleys (which seemed to be caused by expansion) and the way the continents seem to "fit" into each other. However, in the early twentieth century, scientists found that the earth's interior is actually being heated up by the decay of naturally-occurring radioactive elements. They revised the theories and decided that the expanding earth would break the continents apart.

The problem with the expanding earth idea, however, was that it didn't account for the formation of mountains, as the earlier contracting-earth theory did. What was needed was a theory that could account for mountains, rift valleys, and the corresponding shape of the continents. Enter plate tectonics. The originator of this revolutionary concept was a German meteorologist named Alfred Wegener. (His work did a great deal to clear the name of meteorologists, who, even if they can't predict the weather tomorrow, are responsible for telling us what was happening a few million years ago.) In 1910, Wegener developed the hypothesis of "continental drift," where the original block of land, which he called Pangaea ("All Earth"), broke into pieces and went floating around like ice on a pond. Evidence from matching coastlines, geology, fossils, and the ancient magnetism found in the rocks seemed to support the idea that the continents are indeed wandering around and crashing into each other at a slow, but majestic pace.

How does it happen? To understand why the continents drift around, we have to dig deep. Very deep. We'll start from the outside, at the familiar, rocky crust. Actually, the earth's crust comes in two types, the continental crust that you walk around on every day, which is about 30 miles thick and made mostly of granite, and the oceanic crust, which is only about 5 miles thick and made up mostly of basalt. Beneath the crust is the mantle, which is composed of a different collection of rocks, although the exact composition is unknown. And finally, at the center is the core, which is made not of rock at all, but of an extremely hot iron-nickel mixture. (It's about 9,000 degrees Fahrenheit down there.)

Geologists consider the crust and the outer part of the mantle to be the *lithosphere,* which is cold, at least compared to the core. The part of the mantle right under the lithosphere is hot and malleable. The mantle layer below that is the *asthenosphere*, which is so hot that it's close to melting. That makes it weak and unstable. The hot layers below the thin lithosphere have caused it to crack into six large pieces or tectonic plates, which are constantly moving around, a condition called *isostasy*. (There are also quite a few smaller fragments.) So much for the feeling of standing on solid ground. The actual driving force of continent movement is convection, a special kind of heat transfer that also causes hot air to rise. Convection is familiar to most people relating to oven technology, the hot air currents rising from a hot stove or oven. Near molten rock can also have convection currents which cause the movement, not of air, but of the earth's crust itself.

The floating continental plates can move away from each other, grind up against each other, or crash into each other. If they split apart, we get rift valleys like the one in Africa. If two continental plates head-bang into each other—we have a collision zone. Something has to give. And that is how mountains are made, including the Himalayan areas that are home to the world's best rubies.

At the same time, other tectonic plates were crashing in South America and making the world's most perfect emeralds. Thus the fabulous emeralds of Colombia, the matchless rubies of Burma, and the mighty Himalayas were simultaneously born. Although the continents didn't move very fast (only a couple of inches per year), it was enough to create quite a stir, geologically speaking. The famous, thousand mile long Irrawaddy River, which is born in Tibet and empties into the Bengal Bay, is at a "suture zone" between India and the tectonic plates of Southeast Asia. (This same river is the famous "road to Mandalay," upon which the whole economy of Burma at one time depended.) These are the kinds of places, all over the world, where minerals tend to form. That is because as the tectonic plate sinks, molten rocks are pushed out and slowly cooled, making an ideal environment for gem formation.

Oddly enough, despite their vast commercial importance, very little geological research has gone into analysis of these fabulous deposits. However, that has begun to change, due to a special characteristic these stones have, which has interested scientists. Normally, the major constituents of ruby—aluminum, chromium, and vanadium—don't appear in this environment, so researchers were curious. It is known that rubies form at high temperatures—somewhere between 620 and 670 degrees C. By dating minerals that developed along with rubies, such as mica and zircon, it seemed clear that rubies were formed at various times between 40 million and 5 million years ago. (This means that by the time rubies were forming, all the dinosaurs were already dead.)

When scientists studied the liquid inclusions regularly trapped within rubies, they found "feeder fluids" containing salts and carbon dioxide, which came from the high-temperature solution of salts contained in evaporite-bearing beds, found in rich, clay-like, impure marble deposits in Central and Southeast Asia. (Evaporites are pretty much what they sound like they are—what's left over when sea water evaporates.) All the exciting tectonic action set the fluids in motion; as they interacted with the marble, chemical actions occurred that freed the aluminum, chromium, and other ruby-constituents. At first, scientists had thought that these elements were present only in trace amounts; it was determined, however, that when mobile, they occurred in sufficient amounts to create very pure rubies. This new model shows rubies developing deep in the heart of marble, resulting from the involvement of salts and mineralizing fluid of metamorphic origin. The new model suggests that the presence of evaporites is a critical element in explaining these ruby mineralizations, and may give geologists a clue as to where more may be found.

Many of these ideas, however, are somewhat speculative, and given the political situation in Burma (now officially called Pyidaungsu Myanmar Naingngandau by its government) it may be sometime before Western scientists are allowed in the country to do further research on the origin of rubies.

3

The Heart of the Matter and a Matter of the Heart

Talking about the birth of rubies, or even copying out the ruby's chemical formula, tells us little about what a ruby really is, just as a musical score gives little idea to the average person about what a symphony will sound like. For a truer picture, we need to look deeply into the heart of the stone.

It should be said that certain characteristics are dependent on the geographical origin of the stone. Sri Lankan stones, for instance, have somewhat different features than those from Burma. Because the buying public is most familiar with rubies from Burma, these stones often command a higher price than a similar or even better stone from another area, much as Colombian emeralds are automatically assumed to be the best, even when they are not. This false perception makes it possible for an astute buyer to get an excellent "non-Burmese" ruby at a good price.

RUBY STRUCTURE

Like all gemstones other than obsidian and opals, rubies are crystals. And while most of us think of a crystal as something that is "clear," the scientific definition is a little different. A crystal is a mineral that has a regular and repeating three-dimensional pattern. (A crystal pattern repeats about 100 million times per centimeter.) That makes it different from glass, in which the atoms are scattered in a rather amorphous, random way. (If you want to get technical, glass is simply frozen liquid. Obsidian is a rare example of gem material that is glasslike rather than crystalline.) The internal structure of a crystal makes it "grow" in certain specific ways that have little to do with how the gem is finally cut and faceted. Rubies may be cut in a variety of ways, but the internal structure of each ruby is always the same.

Crystals are divided into several "systems" according to the minimum symmetry of their faces. Rubies all belong to what is known as the "hexagonal" system of crystals. (Some texts refer to the ruby's crystal system as "trigonal," but most authorities regard the trigonal system as a subclass of the classic hexagonal system.) There are several other such systems.

Hexagonal crystals have four crystallographic or "reference" axes. (The other systems all have only three.) Three of these axes are at 60 degrees to each other, on the same plane and of equal length. These three are called axes a_1, a_2, and a_3. The fourth axis (called the "c axis") is of a different length and runs perpendicular to the other three. Don't ask me what happened to axis b.

Another characteristic of crystals is symmetry. Crystals have both "planes of symmetry" and "axes" of symmetry. A plane of symmetry is an imaginary "mirror" that divides a crystal in such a way that the image on one side of the plane is a mirror-image of the other side. An axis of symmetry is an imaginary line running straight through the crystal and about which the crystal can be "rotated" so that it looks same two, three, four, or six times if it were rotated 360 degrees. Rubies, like other hexagonal crystals, have a vertical "sixfold" axis of symmetry, so that if you rotated the crystal one time around, you'll get three identical "looks." No crystal has more than this. Some kinds of crystals, such as axinite, have no axis of symmetry at all.

Crystals have flat (planar) surfaces, which can grow in many different ways. In 1669 a man named Nicolaus Steno (his real name was Niels Stensen, but he preferred the Latin version) discovered a law (not surprisingly called Steno's law) that determined that the angle between any designated pair of crystal faces is constant and the same for all specimens of any designated mineral no matter what its shape or size. This isn't always easy to determine in most minerals, which often grow in rather cramped quarters, but when allowed to develop in an open environment, the crystal faces become more evident.

A related aspect of crystals concerns their *habit,* which refers to how they grow. This is true even of crystals that do not develop well-formed crystal faces. Some crystals tend to arrange themselves in long, thin strips; others tend to the fat, pudgy side. Typically rubies have a flat, tabular, or prismatic hexagonal habit, with prominent pincoids. (A pincoid is a plane parallel to two of the crystalline axes.) However, ruby's sister stone sapphire tends to form elongated, barrel-shaped crystals. There is nothing simple about corundum!

SPECIFIC GRAVITY

Specific gravity is another way of talking about the relative density (weight per unit volume) of a gem. The first century Roman writer Pliny the Elder was the first known person to use specific gravity to distinguish rubies from other red stones. Other Europeans were slower to catch on, however, and continually touted spinel and other stones of lesser value as rubies. They should have paid more attention to the careful words of the chemist, astronomer, geographer, physicist, and mathematician Abu Arrayan Muhammad ibn Ahmad Al-Biruni (973–1048), an Arabic scholar from Central Asia of great renown. (Arabic names are somewhat complicated. This one means that Al-Biruni was the father of a son named Arrayan and the son of a father named Ahmad.) Al-Biruni determined nearly perfect specific gravity readings for rubies, spinels, and sapphires. He was one of the only people of early times who accurately divided *yaqut* (corundum) from *balkash* (red spinel).

To find the specific gravity of a stone, compare it with the weight of an equal volume of water. This all has to do with Archimedes and a bathtub, as you may recall from school. It is called the hydrostatic principle, but it isn't as tricky as it sounds. You simply weigh the gem in air, then place it in water. The weight of the volume of displaced water equals what the gem would weigh—if it were made of water. There are various machines that can do all this in a jiffy. (There's another way of measuring specific gravity involving something called a pycnometer, using a bottle with a glass stopper pierced by capillary channels, but we don't need to go into that.)

A volume of one cubic centimeter of water will have a weight of 1.00 gram and a specific gravity of 1.00. (The specific gravity of an elephant, say, would be much lower than that of a ruby, since elephants are mostly water.) The specific gravity of a ruby (and all corundum, which does not vary according to color) is 3.9–4.1. This is very high—the specific gravity of a diamond is only 3.5. A ruby is thus *denser* than a diamond. Therefore, a one-carat diamond *looks* bigger than a one-carat ruby if they are both cut in the same way and proportion, even though they weigh the same. (A carat equals 200 milligrams or 0.2 grams.) Carats can be divided into hundreds—each of which is called a "point." The word "carat" is also frequently confused with "karat," a term which indicates the purity of gold. Pure gold is 24 karats. It does not help that "karat" can also be spelled "carat" with same meaning. Then there is "caret," which refers to a diacritical mark indicating an omission in printing, as well as "carrot," a yellow vegetable supposedly good to eat.

STREAK

Minerals also have a property called "streak." If you rub a mineral across a piece of non-glazed porcelain (a streak plate) it will leave a trace of powder. The color of the powder can help identify the mineral. Ruby, even though it is red, leaves a white streak just as the white mineral talc does. In fact, most translucent minerals that come in various colors (resulting from a "foreign" coloring agent not essential to the basic structure) yield a white streak, even if they appear to be another color. The mineral fluorite, for example, can appear as purple, blue, green, or yellow, but it will always leave a white streak. The reason for this is that the chromium that gives ruby its characteristic red color really takes up very little room in the crystal. Light needs a lot of "travel time" to pick up the coloring effects of chromium so that a "red" ruby still behaves as if it were colorless corundum, at least in this respect. This is also why small crystals are often paler than large ones—and by the time we get to a speck of powder, the color effects are no longer visible.

HARDNESS

Valuable gems are usually hard gems, which not only wear better than softer gems, but also take on a higher polish. Hardness is related indirectly to chemical composition and more directly to the arrangement of the constituent atoms and their inner cohesion.

The most common way of measuring hardness is the Mohs scale, although this scale more accurately measures "scratch resistance" than true hardness. The originator of the scale is Friedrich Mohs (1773–1839), a German mineralogist, who developed the idea in 1822. Although born a German, Mohs moved to Austria in 1801, where he did his research. It happened almost by accident, as he landed a job identifying the minerals in a collection of the wealthy banker J. F. van der Nüll, and Mohs decided that developing a "hardness scale" would be useful for his purposes. This collection, by the way, was important enough to be incorporated into the Imperial Mineralogical Collection in 1827.

Mohs arranged 10 minerals and tested them against one another. Here is the scale, finalized in 1822:

1. talc
2. gypsum

3. calcite
4. fluorite
5. apatite
6. orthoclase
7. quartz
8. topaz
9. corundum
10. diamond

Diamond turned out to be the hardest substance (and indeed is the hardest substance ever found), and talc was the softest. Therefore, talc was awarded a 1 on the scale, and diamond received a 10. According to the Mohs scale, corundum (ruby and sapphire) rank at 9, just below diamond, although there is a quite a difference in hardness. Interestingly, also, even though rubies and sapphires are both corundum, sapphires are more scratch resistant than rubies are.

Technically, however, Mohs' results should be described as a "table" rather than a scale, since the numbers assigned to each mineral are not proportionate to their actual scratch resistance. He didn't really go about this in a truly scientific manner, but just picked up ten minerals he had to hand. Mohs' crystal testing set and handwritten notes on his research are still in existence today.

As you might expect, the hardness of minerals on the Mohs scale do not increase by regular amounts. There are other ways to measure hardness, including the Vickers scale, Brinell scale, and Knoop scale, but none of them offers any real advantages over the more familiar Mohs scale. Obviously, any testing for hardness is destructive to the subject mineral, so it is not a wise idea to "test" your ruby by trying to scratch it.

Scratch resistance is understandably important for gemstones, with a hardness of seven or more being most desirable. Even the ancients mentioned "hardness" as an important quality, which was one way of separating a true ruby from a spinel (which they considered respectively superior and inferior rubies).

Toughness and hardness are not exactly the same things. A stone can be tough (hard to break apart) without being very hard (think about bubble gum). Or it can be hard (like diamonds) without being particularly tough. If you hit a diamond correctly it will shatter. Comparing glass and rubber is another example. If you drop a glass ball onto a slate floor, the glass will probably break. A rubber ball will not. Yet glass is harder than rubber. (On the other hand, rubber will deteriorate a lot faster than glass.) Rubies are not only hard and tough, but also resistant to acids. They simply have an enviable combination of toughness and hardness. It is just one more thing that makes them special.

There are better ways to determine the genuineness of a ruby than by its scratch resistance. And worse ones, too. According to Indian folklore, if you drop a ruby in a cup of milk, the red rays will pierce the liquid. Or, if examined in early morning sunlight against a mirror, the stone will throw out its rays on the lower part of the mirror. This shows that it is a very high quality ruby indeed.

CLEAVAGE, FRACTURE, AND GROWTH SURFACES

Gems can be cut apart in two ways: they can cleave or they can fracture. A stone with cleavage will break in a predictable way. Cleavage is related to the "inner cohesion" of the constituent molecules or atoms in the gem. If a stone has good cleavage, and the lapidary knows his business well, a slight tap on the right place may be enough to split the stone along desired lines. That is their *cleavage*. While good cleavage is handy for lapidaries, it is less helpful when a gem with good cleavage falls on a hard surface; it could shatter. Not all minerals have this quality, and those that do have it in varying strengths.

Most break more easily in one direction than in the others. The *direction of cleavage*, as it is called, depends upon the internal arrangement of the crystals, and gems may have more than one direction of cleavage. These cleavage planes are usually parallel, perpendicular, or diagonal to the crystal face. A crystal face is not the same thing as a cleavage plane. A crystal face is a growth surface; a cleavage surface, on the other hand, is a breakage surface. However, both the planes and faces are related to the gem's atomic structure.

Normally, the cleavage of a gem is rated perfect, good, fair, or poor. Materials like diamond, fluorite, and topaz have perfect cleavage, which means that the cleavage is completely smooth. (Cleavage is not really related to hardness, either, if you define hardness as "scratch resistance," which is how it is defined in the mineral world. Many hard materials, such as the aforementioned diamond, cleave quite easily. So do many softer materials, like calcite or fluorite.)

Ruby, however, is not like most gems. It has such a poor cleavage that it is sometimes rated as having "none," although it has what is sometimes referred to as a "rhombic parting." That doesn't mean a ruby doesn't break—it does. But instead of breaking along predetermined cleavage lines, it *fractures* in an uneven or sometimes shell-like (conchoidal) way. The breakage in fractured gems is quite unrelated to the internal atomic structure ("planes of weakness") of the gem.

All gems can fracture as well as cleave, and when they do, a conchoidal fracture is most common, although there are other ways to fracture, too, including so-called hackly fractures (gold), splintery fractures (nephrite), and uneven fractures, seen most in fine grained or massive gems.

Rubies also tend to fracture when they occur as twinned crystals. Twinning is a phenomenon that occurs when some parts of the crystal are rotated or "incorrectly" repeated. At the juncture of the twinned sections, there is a change in the orientation of the lattice (crystal structure) so that the direction of growth is changed. This occurs more commonly in cloudy, ungemlike rubies, but may appear occasionally in the good stuff as well. Twinning can occur because of temperature or pressure changes while, or even after, the crystal is forming. This is one frequently occurring factor that makes rubies liable to fracture. Still, despite the occasional fracture, most rubies are durable gems highly suited to jewelry.

THE MAGIC OF LIGHT

Even the most glorious of rubies cannot shine in total darkness. For the magic to occur, light must do its thing: reflect, refract, or diffract. The magic of any gem is dependent upon the magic of the light that gives it life and fire. Gems are complex things and handle light in complex ways. Light doesn't just uneventfully flow through windows as it does through glass, or simply bounce back as from a black-hearted mirror. Instead it dances impatiently, refracts, and reflects. It comes alive along with the gem. In a weird way, a gem is a crystal cage that traps the light and makes it fight to escape.

In reflection, the light bounces back at the same angle that it hit the stone. So if a ray of light strikes a completely reflective object like a mirror at 45 degrees, it will bounce back at 45 degrees. Mirrors can *only* reflect. None of the light gets past the surface. *The angle of reflection is equal to the angle change of incidence.* In other words, reflection is symmetrical.

Reflected light is mostly what gives a gemstone its particular *luster.* Two minerals of the same color can vary significantly in luster. Most rubies have a vitreous, or glass-like, shining luster, although some gleam with the adamantine brilliance of diamond. Other vitreous stones include sapphire, emeralds, spinels, and aquamarines. Other gems may have a silky (satin spar gypsum), pearly (moonstones and some organics), waxy (turquoise and some organics), greasy (jadeite), splendent or mirror-like (mica), metallic (hematite and precious metals), resinous (organic gems like amber), or ada-mantine (diamond) luster. How bright that luster is depends first upon the condition of the stone, and second upon the degree to which it is polished, although the gem's refractive index plays a part in luster as well (see below). One type of luster is not better than another—it is merely a matter of personal preference. In addition, some stones, like garnets, have a variety of different lusters.

When light enters a gemstone, some of it bounces back at the surface instead of going into the depths of the crystal. In contrast, mirrors, which can *only* reflect, are totally black at heart. Refraction is different from reflection. In refraction, the light doesn't bounce directly off the object but actually enters into it, becoming in some mysterious way almost part of the object. And as it enters the gem, it bends. You have probably observed this firsthand, seeing an oar "bend" in the water. That is refraction. Why does this happen? It happens because light does not travel at the same speed through every medium. It goes more slowly through air than through a vacuum, even more slowly through water, and almost creeps (relatively speaking) through solids like quartz, rubies, or diamonds. The speed of light equals its wavelength times its frequency. And while its frequency remains constant through various substances, its wavelength changes at the interface between different substances (such as air and rubies), and this causes the light to bend.

The angle at which light decides to *reflect*, which means to bounce back at the same angle it hits (the so-called angle of incidence), rather than refract is called the *critical angle.* If a ray hits a completely reflective object like a mirror at 45 degrees, it will bounce off it at 45 degrees. The critical angle varies with different substances. With polished gemstones, some light is reflected back, but the rest enters the stone and is refracted. For gem cutters to make the most effective cut, it's vital for them to know the critical angle for each material they work with.

The so-called "refractive index" expresses the mathematical relationship between the angle at which the light strikes the gem and the angle at which it bounces away. There's a mathematical formula for this behavior defined as the ratio of the velocity of light in a vacuum (although for practical reasons, we usually just use air) and in a substance. It is measured in terms of the ratio of the sines of the angle of incidence and the angle of refraction at the interface between the two media. However, that is far too many prepo-sitional phrases to be placed in any one sentence to make complete sense, and is possibly the reason why most people have no idea about what a refractive index is—or is not.

In any case, the refractive index of rubies is 1.76–1.77; the range indicates the difference between the minimum and maximum values. The higher the refractive index of a substance, the greater the bend in the light. (Scientists use a special refractometer designed for gemstones to measure refraction.)

Gemstones belonging to the hexagonal system, like rubies, are "doubly refractive" or "birefringent," which means that as the light enters the polished stone it is split into two separate rays. Each ray is refracted to a different degree. The birefringence of ruby is 0.008. However, red spinel, which is often mistaken for ruby, is only singly refractive (as is garnet). This is one easy way to tell the difference between the two. Another common ruby imitator, glass, is also singly refractive.

In addition, some particular facets may seem lighter or darker in tone; this is a result of the ruby's pleochroism. Pleochroism, which means "different colors," simply refers to the way a stone's color changes when it is turned. Typically, rubies look yellow/orange or brick-red in one direction, and deep carmine/purplish red in the other. Ruby is thus considered a *pleochroic* stone. (All members of the hexagonal, tetragonal, and trigonal crystal systems share this property, showing two different colors, making them dichroic.) Blue sapphires show up greenish blue in one direction and violet-blue in the other. Members of the orthorhombic, monoclinic, or triclinic systems show three colors, making them trichroic, while amorphous and cubic system stones have only one. To see pleochroism with your naked eye, you need to hold the stone, remember the color, and then turn it. In case your memory is not very good, you could invest in a dichroscope that will show you both colors at the same time. Pleochroism is strongest when the table facet lies parallel to the *c* axis. One bad effect of pleochroism in rubies is called "bleeding," a phenomenon that shows up mostly in lighter, less desirable stones. If you take a ruby from natural light to incandescent light, it will turn pinkish. The more apparent this effect, the less valuable the stone is.

COLOR: THE PIGEON BLOOD FACTOR

Light does more than make the surface of a ruby glitter and dance. It also reveals its color, the single most important gauge of value in a ruby. A stone of good color outranks a larger or even less flawed stone of poor color. Color is really nothing more than a form of light-energy. For color to exist, there must be light. In a dark cave the reddest ruby literally has no color at all. The color is born with the light.

Color is partly dependent on how the gem absorbs or reflects light. When white light (which contains the entire spectrum of colors) strikes a colored gem, only some of its colors are absorbed. This is called "preferential absorption." The colors that are *not* absorbed are what give the stone its color. A red gem like ruby has soaked up all the blue and green light, leaving the red. Here's how it works. A beam of light strikes an electron within the gem. If that light has just the right amount of energy to kick the affected electron into a higher orbit, the light is "used up" doing its job, so to speak. When the electron falls back to its original orbit, it gives off the extra energy as heat. And this goes on as long as there is light upon the stone. Which particular color of light an electron will absorb depends on the energy difference between orbits, and that in turn depends on the structure of the mineral involved. Of all the colors, red is the "least energetic" and violet is the most.

If a stone absorbs all colors, it will appear black like onyx; if it absorbs none, it is clear, like a diamond. It has only been comparatively recently, with the development of the "brilliant cut," which can take full effect of white light, that colorless stones have been fashionable. Before that development, all valuable stones were stones of color.

As diamond is the ultimate colorless gem, emerald the ultimate green, and sapphire the ultimate blue, so ruby is the definitive red. Red in itself is the most mystical of colors, even scientifically speaking, if one can allow a bit of mysticism to enliven science. It has the longest wavelength of all visible colors and is the nearest visible wavelength to infrared (which creates the feeling of heat). In fact, when vision is restored to a person after a period of darkness, blindness, or trance, red is the first color seen. And of course, red is the color of blood with its own myriad associations of power, passion, and violence. In some languages the words for "red" and "blood" are the same.

All red stones are not equal. In point of fact, every species of gem has its own specific color, even though they may appear the same to the naked eye. The color of a fine ruby is nothing like that of garnet or morganite or kunzite, which are all very nice stones in their own way, of course. But they are not rubies. A spectroscope will soon sort out a ruby from a red garnet or a red spinel by identifying the color spectrum unique to each gem. Incidentally, it is a myth that garnets and spinels are always red. One variety of garnet, the demantoid garnet, is green and is very valuable.

Even though the most noticeable and commercially important aspect of the ruby is its matchless color, it is always a little surprising to recall that this color is not part of the intrinsic structure of the stone, but a foreign intruder. This is common in the gem world. Gems whose color results from an impurity (not part of the essential chemical structure) are called "allochromatic"; those gems whose color is *part of their essential structure* are called "idiochromatic." Peridot and malachite are examples of idiochromatic gems.

Malachite, for instance, has copper as part of its essential structure, and copper absorbs all colors of light encountering its atoms except green. Peridot, another green stone, gets its color from the iron that is part of its own essential structure.

On the other hand, the ruby, like the emerald, is an *allochromatic* gem. As mentioned, it gets its color from a few stray atoms of chromium that have sneaked into its crystal lattice in the form of an ultrafine vapor. Chromium is relatively abundant (as such things go) in the crust of the earth. In its natural state, it is a brittle, hard, corrosion-resistant, steel-gray metal. In nature, it is never found alone, however, but is always combined with other elements, especially oxygen.

Chromium was first isolated in 1797 by Nicolas-Louis Vauquelin (1763–1829), a French chemist. He extracted it from Siberian "red lead" or crocoite ($PbCrO_4$). Today chromium is primarily obtained through another compound called chromite. The name chromium comes from the Greek word for "color," fittingly enough. Vauquelin is also the one who discovered that his newfound mineral also imparted the green color to emeralds. Chromium therefore makes rubies red and emeralds green. The reason for this chameleon effect lies not with the chromium per se, but with the different arrangement of atoms in beryl (emerald) and corundum (ruby) that influence the position of electron orbits in different ways.

Allochromatic stones are usually much more capable of being "enhanced" by heat treatment than are idiochromatic gems, although there are plenty of exceptions. It works with rubies, for example, but not with emeralds.

Color is not a single entity. It has several aspects, including hue, saturation (intensity), and tone (value or darkness). *Hue* is a synonym for what we normally call "color" in casual conversation. A pure hue is a rare thing in gemstones, although it is usually very desirable. Red, blue, and yellow are all hues. (Black and white are not properly considered hues.) Brown covers a range of hues of low saturation and usually high value. Corundum comes in a variety of primary hues, including red, blue, yellow, and pink. The hue of rubies is, of course, red. While the standard for ruby is medium clear red, in actual fact you can find rubies with definite brown or pink shadings. At one time the Burmese called these stones "blood-drops from Mother Earth." In the estimation of many, the best stones are pure red—red with no blue or brown notes. Very light or dark shades are less valuable, although they may appeal to certain buyers, who are then in luck, as such stones are definitely cheaper. In any case, the true, perfect, unadulterated hue we think of as true "ruby red" is very rare.

In addition to their primary hue, most stones have a "secondary" hue, usually of some related color, which appears as the stone is turned in various directions (pleochroism). In rubies, red is the primary hue; the secondary hue can be various shades of pink, violet, or orange. However, most experts will agree that a "fine" ruby should have at least 85 percent "red" as the primary hue. This makes them a pure, bright, medium red. Stones with less red than that are usually given the lower grade of "commercial."

The second element of color is *saturation*. Saturation measures the vibrancy, richness, or intensity of a hue. A pure color for a ruby is one with no brown (or less frequently, gray) mixed within it. If you take a drop of pure blue dye and drop it in a glass of clear water and then continually add more blue drops, the saturation increases, although the hue does not change. It's still blue. The easy way to remember all this is that saturation simply means more or less of a particular color. In the world of rubies, Burmese stones have the best saturation. In fact, these stones have the richest saturation of any gem with the exception of tourmaline from Paraiba, Brazil. The strong fluorescence (more about fluorescence later) of most rubies vastly increases the color saturation of the stone. It is a glorious accident of nature that rubies possess both red color and red fluorescence.

A related term is *tone* (or value), which is the darkness or lightness of a color. Gemstone tone is evaluated as "light," "medium-light," "medium," "medium-dark," and "dark." White has no darkness, while black has all darkness. Tone is determined by the addition of another color to lighten, darken, or brown it out. Some colors are more naturally dark than others even at equal saturations. The best rubies have a vivid pure red hue with a medium tone, and with no modifying brown tone. The desirability of slight modifications by orange or purple are a matter of taste.

If you have glass of highly saturated red water and start adding blue, the water will grow darker in tone even though its saturation has not changed. In the practical world, it is very hard to distinguish saturation from tone in gemstones. In gems, red reaches its optimum saturation, or its "gamut limit," at between 75 and 80 percent tone. If the tonal value is below 60 percent the stone looks washed out. If it is over 80 percent, it looks too dark.

The ancient Greek writers took the lightness and darkness of a stone quite seriously, for they not only thought that the lighter colored or pink stones were "female" and the darker and star stones "male," but also that the two could unite to produce offspring! Even today, some crystal healers believe that best results will be obtained from obtaining a "mated pair." This idea persisted at least into the sixteenth century, when it was asserted that a couple of diamonds belonging to a "noblewoman of the house of Luxembourg" had actually managed to produce a litter.

In faceted gems, the true hue, saturation, and tone are best seen and evaluated in the so-called "internal luster" of the stone. Light entering the gem through the crown or table facet reflects off the pavilion facets internally then returns to the eye; light entering the gem through a pavilion facet reflects off the opposite pavilion facet and then comes to the eye.

Since 1989, the International Colored Gemstone Association (ICA) defines *all* corundum with a red or pink color as ruby. Not every gem association agrees with this assessment, and there's an ongoing battle as to what makes a ruby a ruby. (The same battle is being fought over the difference between "emeralds" and "green beryls.") There's an old joke in the jewelry business that goes something like this: "Whether a stone is a ruby or a pink sapphire depends on whether you're the buyer or the seller." It also depends upon *when* you were doing the buying or selling as fashion tastes and consequent value change over time. A similar dispute concerns how red a "pink sapphire" has to be before it can cross the magic threshold into true rubydom. After all, there is more money in rubies than in sapphires of any color. However, intensely colored pink sapphires (or rubies) are worth a good deal of money.

The term "pink sapphire" did not appear in the literature until the beginning of the twentieth century. These stones used to be called pink rubies, and because of their comparative rarity, have sometimes been valued as much as rubies of the more standard hue. Due to a quirk in the English language, pink and red are sometimes considered two different colors, when in reality red is just a more saturated pink. They are both the same hue. The word "pink" in fact did not enter the English language until the eighteenth century. The color is named after the flower (not the other way around), a relative of the carnation. The flowers themselves didn't show up in Great Britain until the sixteenth century. The color we call pink was previously referred to (if at all) as "rose."

Notice the same terminology confusion does not occur in relation to blue sapphires—all of which are termed "blue," no matter how light or dark they are. Blue is blue in English, but red is sometimes pink. This is all very illogical, but there's not much anyone can do about it now, except to stop calling pink corundum pink sapphire. I predict this will not happen.

Some experts divide the color of pink/red corundum into color subgroups. "Hot electric pink" stones are to be classified as pink sapphires. Another group is classified as "electric magenta." While many rubies seem to have this purple/magenta tinge to them, the rule is that the more apparent the purple, the less valuable the stone.

In this scheme, only a third sub-group, stones of a "stoplight red or cherry candy" hue, would be considered true rubies. It appears that modern gemologists are not the first to create strange metaphors for the various colors of ruby. The most widely used term for this color is "pigeon blood" (*padamya nyunt*), an obscure and ridiculously romanticized phrase that has sparked a good deal of controversy, most of it unnecessary. It should be noted that pigeons did not get the rather unsavory reputation they now carry until the middle of the twentieth century. Previously, they were more respected, and the term "pigeon blood" was one of honor.

No one is really sure who first used the term—or why. One legend tells that a gem dealer wanted to get to the bottom of the story by checking out what pigeon blood really looks like. The story goes that he traveled to Rangoon, where he bought twelve Burmese pigeons, and then hired a Buddhist priest to slaughter them. On a certain day, at high noon, he went to the beach with the priest and the pigeons in tow. With his sword, the priest decapitated the birds one by one, then waited three minutes and poured the blood of the pigeons into his palm. The color? Bright pink.

There is so much wrong with this story one does not know quite how to begin. In the first place, there are no Buddhist "priests" except in Japan. Second, Buddhists of any sort are not in the habit of slaughtering animals, especially just to find out what their blood looks like. Third, why would one need twelve pigeons? Or to kill them at noon? Or even to go to Rangoon? One supposes the blood of Rangoon pigeons is the same color as that of any other pigeon—which is red. Possibly the gem dealer had a bundle of pink sapphires he wished to unload and thought by this underhanded tactic that he might persuade people to give a higher price.

Leaving Buddhist priests aside for the moment, it is known that pre-Buddhist Burma was rife with a form of animism known as Nat, and the original inhabitants frequently killed pigeons and chickens to honor the forces of the other world. (It is undoubtedly still done in remoter areas of the country.) Possibly the term comes from this practice. However, the animals are killed not with a sword but by twisting their necks. As a result, tiny droplets of blood dribble out of the nostrils onto the animal's beak. Perhaps that is the root of the story.

Another approach to comprehending the incomprehensible pigeon blood story is the suggestion that the color refers to a pigeon's bright red eye. If this were the case, though, it would seem that the color should be called "pigeon's eye red," a phrase which if anything is even more unfortunate than "pigeon's blood." Calling a pigeon a "dove" helps the phraseology somewhat, but doesn't really get to the heart of the matter. One supposes that dove's blood is much the same color as pigeon's blood. They are after all, closely related birds. (There has been a halfhearted effort on the part of some to substitute the term "beef blood" for pigeon blood, an attempt one hopes is doomed to failure.)

In a modern approach to examining the color question, one James B. Nelson sought assistance from the London Zoo, as he reported in a 1985 article in the *Journal of Gemmology* (XIX, 7). He wrote: "Their Research Department were quick to oblige and sent a specimen of fresh, lysed, aerated, pigeon's blood. A sample was promptly spectrophotometered. . . ." One can only hope no pigeons were actually killed in this ridiculous endeavor.

The term has even entered literature. In 1955, Joseph Kessel penned a travelogue/novel called *The Valley of Rubies* featuring a couple of ruby merchants who travel to Mogok to see the true pigeon's-blood (or in Burmese *Ko Dwei*, a word that may ultimately be of Chinese or Arabic derivation) stone. The very words "pigeon's-blood" were "like some mystic incantation, some magic password," he enthuses. The sad truth is that some people cannot appreciate a poetic metaphor, which is simply that the color of the best ruby glows like a drop of the reddest blood upon the neck of the whitest dove. That's really all there is to it.

This phrase does little to actually describe the color, however, and the exact shade of red that earns the epithet remains in dispute, with some sources claiming the proper color has a tinge of purple, while others insist that pigeon-blood stones are free from any color other than red, enhanced, perhaps, by its natural fluorescence. (There is an ancient tale that claims Noah used fluorescent rubies to guide the ark.) More tellingly, an anonymous nineteenth century Burmese trader is reported to have remarked, "Asking to see the pigeon's blood is like asking to see the face of God." Be that as it may, gemologist Richard W. Wise, in his *Secrets of the Gem Trade*, reported that he was privileged to glimpse a true pigeon's blood ruby, during a trip to Burma. He compared its color to "a rich tomato sauce that has simmered for hours on the stove," with a touch of blue thrown in. Perhaps not the happiest of metaphors, as "tomato sauce red" seems to lack a little mystique. Besides, spaghetti sauce with a blue tinge is a singularly unappealing idea.

The Burmese color-evaluation system did not stop with "pigeon's blood." The second-best color is called "rabbit's blood" or *yeong-twe.* It is a slightly darker and edges towards blue. The third best is a deep hot pink termed *bho-kyaik,* which literally means "preference of the British," possibly referring to the penchant of the British Mogok gem dealer A. C. D. Pain for this shade. (A. C. D. Pain was so important that the mineral painite was named after him.)

Fourth down on the list is *leh-kow-seet* (literally "bracelet-quality" ruby), a light pink. The final and lowest quality is the dull, overly dark *ka-la-ngoh,* which literally means "crying-Indian quality." Most of these darker rubies were in fact sold in Bombay (now Mumbai) or Madras (Chennai), India, and some suggest that *ka-la-ngoh* stones were so dark that even Indians would cry out in despair when confronted with this unlucky color. But people are just guessing.

We do know that the best ancient rubies had the same qualities as the best modern ones do: smooth, red, bright, lustrous, flawless, and transparent. True flawless rubies, however, are rare. (Flawless diamonds are much more common.)

FLUORESCENCE

Fluorescence is a very important property of many gemstones. It is similar in many ways to what we just discussed about light, but fluorescence is triggered by ultraviolet light. Ultraviolet light is invisible to us and packs a lot more energy than ordinary visible light. When ultraviolet light hits an atom, it can push electrons into higher orbits just like regular light; however, in this case, when the electrons fall back to their accustomed orbits, they give off not heat but visible light—they shine in the dark! The color they shine depends on how much excess energy they have picked up. Burmese rubies fluoresce strongly to long wave (366 nanometer [nm]) ultraviolet radiation and less strongly to shortwave radiation. (The visible spectrum of sunlight ranges from 740 nm to 450 nm.)

Rubies' fluorescence is apparent in both artificial light and in some cases even in daylight, making the gem appear truly radiant. The fact that many Burmese rubies actually fluoresce to *visible* light is rather unusual, although some spinels share this highly sought-after property. The ancient Burmese considered this feature supernatural—and in some cases a product of witchcraft. In reality, of course, there is nothing unnatural about it, but it is strange. (Fluorescence also helps mask the "extinction" phenomenon in cut rubies, one factor that tends to dull their effect.)

Most rubies fluoresce a strong carmine red, which serves to enhance the stone's natural red color. Red rubies turn even redder—and purplish ones more nearly approach the ideal red color in good lighting conditions. At one time it was believed that by looking in the strange, fiery fluorescence of Burmese rubies, one could see dragons and other mystical beasts.

Not all rubies possess the same quality of fluorescence, however. While Burmese and Vietnamese rubies may fluoresce even in daylight, giving them an incredible inner glow, rubies from Thailand will not. This is yet another reason why Burmese stones are considered so valuable. So even though some Thai-Cambodian rubies have technically better color than the Burmese variety, the fact that they don't fluoresce in visible light has diminished their overall appearance and hence value.

GROWTH ZONING

A good stone shows uniform color when looked at from the face-up position. However, many rubies (and sapphires) show color zoning or, as it is sometimes known, color banding or growth zoning. These are variations in color that occur throughout the gem, or sometimes as bands of color alternating with bands of colorlessness. They usually occur in growth zones as straight bands, but may also show up as spots. Rubies are generally cut so to appear uniformly red when viewed looking downward at the face; the banding is usually apparent when the stone is viewed from the side or by placing it face down on a white surface. It should be said that the effort to avoid the banding effect may force the stone out of proportion, or result in odd facet patterns. The entire ruby may be an unusual shape. It is generally considered that almost any distortion of the stone is allowable to enhance the color, but it is still not ideal.

In rubies from Jagdalek, Afghanistan, this zoning is sharp and narrow, forming a typical hexagonal pattern when viewed parallel to the c axis. These rubies often show little spots of blue as well. Sometimes these blue areas are hexagonal; other times they are narrow ribbons. But in all cases, there's a very distinct line between red and blue zones. Vietnamese and Burmese rubies from Mong Hsu show similar color zoning, and are often heavily fractured as well. Mong Hsu stones even produce a rare anomaly—a ruby with a pure blue sapphire core.

The best gems have a high degree of *color coverage*, no matter what angle the gem is viewed from.

INCLUSIONS

Even the earliest gemologists knew that a perfect ruby was a rare find. The eleventh century gemologist Al-Biruni, for example, wrote about common flaws found in the stone. He remarks:

Among the blemishes of the ruby which Al-Kindi [an earlier writer] has mentioned is the inner strain which, if too conspicuous and deep, cannot be removed. The other is the *khalt-i-hijarah* (admixture of stones) which is called *hurmulliyat*.

Hurmal (harmal or white rue) is white. In Persian it is called *kunjdah*. Another blemish is that of rim, i.e., a kind of dross that is like earth. Still another is that of a perforation which detracts from its clarity and transparency. This appears in the form of a crack which results from the collision of a vitreous object which something and the crack is so wide that water may pass through it. It is physical as well as temporary. Variegation in color, e.g., greater in one part and less in the other, is counted as a defect. Cloudiness also deducts from the value of the stone. A pearl-like stain may be present on the stone on any part. This blemish is known as *asin*. If not deep, it would disappear on rubbing the stone. There is no other way in which to do away with this defect, as it is rather deep. (*The Book Most Comprehensive in Knowledge on Precious Stones: Al Beruni's* [sic] *Book on Mineralogy*, Al-Palam Publishing, 1989)

Today we call some of these same flaws color zoning, healed fractures, and included solids. While most inclusions hurt the value of a gemstone, they are geologically interesting nonetheless, for their presence (or absence) tells the history or provenance of a stone. Their appearance is also a very good clue that the stone in question is natural, not created in a laboratory. Rubies from the Sagyin mines (outside the so-called Mogok Stone Tract or main mining district) are noted for few rutile particle inclusions, calcite inclusions, and tubular prisms. Crystals of hematite (iron oxide), ilmenite (iron/titanium oxide), or hematite-ilmenite mixtures may also form, leading to cloudy or silk patterns within the ruby. Rubies from the Nysaseik mine, also outside the Mogok Stone Tract, have inclusions of dolomite, apatite, and mica grains, as well as calcite. True Mogok and Kyatpyin rubies have their own special forms of included calcite, dolomite, apatite, and feldspar.

Inclusions can be gas, solid, or liquid, and can be categorized as *protogenetic* (meaning that the inclusions formed before the host crystal and were "captured" by it as the host crystal grew), *syngenetic* (inclusions that were formed as the same time as the host crystal), and *epigenetic* (inclusions that appeared after the formation of the host crystal and which were trapped into the cracks or cleavages of the host crystal).

Protogenetic inclusions are always solid. Ones that developed long before the ruby itself are typically "etched" or corroded; those that formed immediately prior to the development of the ruby tend to be well-formed. In rubies, typical examples of protogenetic inclusions are apatite and spinel crystals, which were incorporated into the stone under extreme heat and pressure.

Calcite and dolomite can appear as syngenetic inclusions in rubies. These inclusions often form in metamorphic rock such as found in Mogok, Burma. Primary cavities or "negative crystals" may also appear as the crystal is growing rapidly. These primary cavities may be found in all crystals and may be filled with liquid alone (single phase inclusions), liquid and gas or liquid and solid (two-phase inclusions), or liquid, gas, and solid (three-phase inclusions). In some cases, the gas inclusion may actually move around within the cavity, a movement that can be seen under a microscope and which may have been going on for millions upon millions of years.

Epigenetic inclusions may form immediately—or millions of years—after the crystal has stopped growing. Such formations are often called "exsolved" or "unmixed." What happens is this: as the crystal cools, some impurities are excluded from it while others tend to form crystals within the ruby. All exsolved crystals form patterns, and gemologists can often identify the inclusion simply by the pattern it makes. Rutile, which is the microscopic, interlaced needlelike inclusions so characteristic of rubies, is often cited as an example of such epigenetic inclusions (although rutile is sometimes considered to be syngenetic also). Rutile is a titanium oxide mineral, and in fact, is an important source of titanium. (Rubies, of course, aren't the mineral actually mined to make titanium!) Titanium oxide makes a nice white paint and is an ingredient in many whitening toothpastes.

Rutile appears in both ruby and sapphire, and can even form in quartz. Rutile is special in another way. Unlike other inclusions, rutile in rubies is not necessarily considered a defect. In fact, the rutile present in the best rubies, in the right amount, actually enhances the gem's value by adding fire to the stone. As it forms, the rutile tries to migrate out of the ruby, but has to move along narrowly defined spaces—in the end it forms a beautiful spider-web kind of pattern or "silk sheen." Rutile actually has many times the fire of diamond, but this interesting property is masked as it occurs in red, brown, or black colors. The rutile present in many rubies can actually increase color coverage by scattering light into areas of the stone it might not strike otherwise. The scattering gives rubies a soft, warm appearance, quite blood-like, in fact. However, if there is too much rutile, it can cloud the appearance of the ruby by desaturating the red color of the gem and giving the stone a dull, gray appearance.

Another kind of epigenetic formation is a healed fracture. Any time after a crystal has formed, it can crack due to trauma or pressure. Dissolved solids usually leak into the crack, healing or partly healing it, a process that may take millions of years. Eventually the healed crack forms a feathery or fingerprint-like pattern. An unhealed crack in the stone is one of the most dangerous of faults, since it increases the likelihood of damage to the gem. "Healed" fractures are much less harmful. Any sort of crack in the edge of a stone is more dangerous than one in the middle. Even synthetic rubies may have healed fractures, although their microscopic appearance is somewhat different from those of natural stones.

Almost all rubies, even stones of excellent quality, have inclusions, but most of them are not visible to the naked eye. A good ruby should therefore be "eye-clean," which means that to the naked eye it is free of inclusions, although some inclusions may appear under magnification. It is rare to find larger stones that are completely free of inclusions.

THE STAR AND CAT'S-EYE RUBY

Like its sister stone, the sapphire, some rubies display a six (or on occasion twelve) rayed star effect when cut into a cabochon shape, which resembles a blood drop. (The spinel will also form stars.) The technical name for this phenomenon is "asterism," but is famously known as the "star ruby." According to Indian legend, star rubies are always "male" stones. The luminous star is caused by a special orientation of the intersecting needle-like rutile inclusions within the stone, which align themselves at 60 degree angles and reflect light in a special way. They exactly "mimic" the hexagonal habit of the ruby crystal itself. The cabochon (the word comes from Old French and means "head," because of its shape) can be either round or oval, but when oval, the rutile lines follow the long axis of the stone. These star-like formations seem to move around when the gem is rotated, but that is just a trick of the light. For asterism to occur, there must be a sufficient quantity of rutile, and equally important, the needles must be properly arranged. During the 1950s, synthetic star rubies and sapphires were sold under the trade name "Linde Stars." These are still being produced.

The best of these stones have straight "rays" or legs of equal clarity. As with every other variety of ruby, the best stones are Burmese, although they are perhaps more common elsewhere. Burmese star rubies are brilliantly red. The rays can be silver or, rarely, red. Sri Lankan star rubies have a wider color range, from pink to violet-red. (Star rubies are never heated treated, as that would destroy the rutile needles that provide the characteristic "star.") Sri Lanka also produces fine star sapphires in a variety of stunning colors. The rays are usually silver and very strong. These stones have a brilliant quality. Star rubies from Mysore, India, and Vietnam are more opaque, and of inferior purplish color. They are also a lot cheaper.

The largest ruby in the world, the Eminent Star Ruby, a massive gem of 6,465 carats (almost three pounds), is a star ruby, although it is considered of only mediocre quality at best. In the rough, the stone weighed over 30,000 carats. It is believed to be Indian in origin and is presently owned by Kailash Rawat, of Eminent Gems.

Similar to the star is the "cat's-eye" ruby. This effect, technically known as chatoyancy and best seen under a strong light, occurs when the needle-like inclusions are arranged in parallel. Sapphires also can show a "cat's eye." Again, the stone must be cut in cabochon for the effect to be seen. With the best cat's eye stones, the "eye" appears to open and close when the gem is rotated 360 degrees.

4

Caste, Color, and Cost:
It's All Related

The grading of rubies is an art almost as old as civilization. The ancient Indians, who divided their society into four main castes, did the same for rubies (and diamonds too). I should note, however, that these terms are highly permeable and subjective. In no case can we be certain that any stone assigned to any class is truly a ruby, as the ancients had no really good way of separating rubies from spinels, although they certainly tried.

Grading rubies was traditionally done during the full light of day, since artificial light of any kind tends to darken stones, so that poorer quality light colored or orangey rubies appear more red. Early traders in the Shan district of Burma used a highly polished copper or brass plate that reflected daylight and, it was believed, could be used to separate rubies from the lower-valued spinel.

The highest class in Hindu society was that of Brahmins or priests. These were the only people authorized to preside over the sacrifices that were the heart of the early Vedic religion. The highest "caste" ruby was likewise the "Brahmin" ruby. The Brahmin ruby was usually a pure red color, but not always. A case can also be made for the *padmaraga*, or lotus-colored lily (probably *Nelumbo nuicfera*). *Raga* means "color" and *padma* means "lotus." This Indian red lotus is more deep pink than ruby red. There is also a great deal of confusion here with the Sri Lankan term for a highly prized sapphire: *padradascha* (a term indeed derived from the Sanskrit *padmaraga*, but not referring to precisely the same color). This orangey-red is not usually the ideal color of a ruby, but it makes a truly fine "fancy" sapphire. The reasoning, such as it is, for having pink or red stones representing Brahmins, is that high-caste Brahmins were supposed to have a pinker or redder skin than members of the other castes, whose skin was supposed to prgressively darken the lower down on the social scale they were. In fact, the Sanskrit word for caste was *varna*, literally "color," from which we derive the English word *varnish*.

As mentioned earlier, many Indian sources denote the same color as *padmaraga* or "red as the lotus." Lotuses come in many colors, but the most common variety in Sri Lanka is this orangey-red one. This is not usually the ideal color of a ruby, but it makes a truly fine fancy sapphire. In Buddhist countries the lotus is one of the great symbols of the faith, and any comparison with a lotus is deemed favorable. Every important Buddhist deity is associated in some way with the lotus; he is either sitting on a lotus or holding one in his hand. The most common association is that of divine birth, the lotus rising from the muddy water, then flowering in the pure air.

In Buddhist lore, the red lotus (*kamala*) is the seat of compassion and love, the symbol of the heart's original nature; it is the lotus of Avalokitesvara (Tara), the Buddha of infinite compassion. However, it is the pink lotus that is the most treasured, and the symbol of Gautama, the historical Buddha. The pink hue is also referred to as "sunset," at least according to the *Rayanaparikkha,* a medieval Prakrit text on gemology. Sugared lotus seeds are eaten even today, and it is said that even after a thousand years, the seeds can be re-planted.

One of the curious aspects of the Sri Lankan *padradascha* is that the finest examples derive their color not from saturation of a pure hue. Indeed, the color is a mixture of hues. This gem is considered, in Indian astrology, to be a "feminine gem," probably because of its pinkish color. Its power was such that even an ignorant person living a sinful life would be saved by the power of the stone. However, we should note that this color corundum should probably be termed "pink sapphire" rather than ruby. It should also be said that some sources consider the Brahmin ruby to have a pure true red color, none of this sunset/lotus stuff.

Whatever its color, the Brahmin ruby has infinite protective powers for its owner, who is safe from any disease or disaster, particularly from the Evil Eye. But there's a caveat. If the Brahmin ruby is allowed to be contaminated by the touch of a lesser stone, its powers will fade or even completely vanish. In the same way, ancient Brahmins kept themselves aloof from the other, lower castes. The great glory of the Brahmin ruby was that its power was perfect, just as the Brahmins were supposed to have perfect communication with the gods.

The second caste in Indian culture was the Kshatriya, or warrior/leader class. The ruby belonging to this class was the rubicelle, a yellow/orange/brown ruby or spinel, whose color in some odd way was supposed to correspond to the coloring of Kshatriya caste members.

The third caste was that of the Vaishya, the farmer/merchant/artisan caste who once formed the largest part of the community, before the Shudra, or servant class, was added. They were to support the community through farming and other independent labor. Their "ruby" was probably a spinel of some description.

Fourth and finally came the Shudra caste, the laborers. Their ruby was the balas ruby, which like all the above terms has a variety of definitions, but in any case certainly referred to the lower-quality stones, including spinels. In point of fact, it is unlikely that members of the ancient Shudra caste were able to afford jewels of any kind. As the caste system developed, it became a truly "social system" which in many cases had nothing to do with the careers or even the wealth of the disparate members. Today, it is certainly possible for a Shudra to have more money than a Brahmin, for instance. However, the delicate social distinctions remain even now, especially in rural areas, and most Brahmins would not like to share a meal with a member of a lower caste. (The people called "untouchables" are an offshoot of the Shudra caste.)

The confusion between various shades of red and between rubies and spinels obviously created problems, although more for Europeans in later times than for the ancient Indians and Burmese, who graded mostly, although not entirely, by color.

The word *spinel* is of uncertain derivation. It may stem from the Greek word for spark, the Latin word for point, or the Italian word for thorn. (Spinel crystals tend to be pointy.) As we can see, rubies and spinels were regularly mixed up by the ancients. Like ruby, spinel can fluoresce in natural light—a rather uncommon phenomenon. However, spinel is only singly refractive, not doubly refractive like the ruby.

Red spinels are so lustrous and attractive in and of themselves that the Burmese natives who first found them dubbed them *anyan-nat-thwe*, which means "polished by the spirits." They are extremely hard to distinguish from the more valuable ruby by eye alone. However, true rubies could not hide from gem-cutters, who knew by the exquisite hardness of the stone (and consequent difficulty in cutting) that they were dealing with something quite different from spinels. What they may have told the public was another matter.

It wasn't until 1783 that the clear crystallographic differences between the two were discovered by a scientist named Jean Baptiste Louis Rome de Lisle (1736–1790), although, as mentioned earlier, Pliny the Elder and al-Biruni had also discovered the differences by using specific gravity. De Lisle was taken captive by the British in India in 1761 for no very clear reason (although he was secretary of a company of artillery), and was held in prison there for three years. It was after that that he gained fame as crystallographer, but may have developed his interest in rubies in India.

Although spinel is not a ruby, it is still a well-respected stone, valued not only for its beauty as a jewel, but in former times for its magical powers, just like the ruby. At one time it was believed, for instance, that a spinel could somehow detect if an apparently ordinary person was in reality a supernatural being.

Today, the grading of rubies and other colored gemstones is a highly contentious matter in the gemstone world. Everyone is familiar with the De Beers cartel for diamonds, which manages to artificially support diamond prices around the world as well as pour in millions of dollars worth of advertising ($90 million per year in the United States alone) promoting this stone. Until recently, rubies have not enjoyed such attention, as there is no single monopoly to control their production and sale. However, in 1997 the American Gem Trade Association (AGTA) and the International Colored Gemstone Association (ICA) began to promote rubies heavily, with an initial investment of $400,000. Unfortunately for them, they have so far failed to come up with a slogan as memorable as "Diamonds are Forever," and consumer interest in rubies remains comparatively low. (The highest price ever paid for a ruby necklace was one sold by Sotheby's in 1989. It went to Van Cleef & Arpels for $3,080,000. It has diamonds on it too.)

Rubies are also more common than they should be, at least if we are considering human rights. While Burmese goods are generally not permitted to enter this country because of the human right violations of the present regime, U.S. Customs has spotted a "loophole" in the law that permits the importation of rubies (the Burmese Freedom and Democracy Act of 2003 and extended in 2006 for a subsequent three years). In December 2004 it declared any gemstone of Burmese origin that is *cut* outside Burma (and almost all are) has been "significantly transformed" and is therefore not considered an item of Burmese origin, and so is not regarded as a sanctioned item. Thus, with a wink and nod, the coveted Burmese rubies, which command a higher price than Thai or Sri Lankan stones of equal quality, are once again freely available on the market. All of this, of course, supports the brutal Burmese government and drives down ruby prices. The prices for colored stones have always been less stable and less controlled than prices for diamonds, partly because of the poverty and political instability of the producing countries (making the marketing of them more dangerous and erratic), and partly because the mining operation for most of these stones remains fairly small-scale compared with diamonds. Unlike rubies, diamonds are graded and labeled in a mechanistic, objective way that requires no particular skill or training, and which is easy for the public to understand. A similar grading system for rubies would have the advantage of increasing consumer confidence about the quality of the stones they buy.

The ancient Romans prized the ruby above the diamond, calling this gem "a flower among stones." For the Greeks it was the "mother of all gems." And back in 1560 Benvenuto Cellini declared that the price of ruby was eight times that of diamond. Of course, that was before the brilliant cut was developed for the diamond, which significantly enhanced its looks.

The ruby has always been, and remains today, the world's most precious gemstone, if we exclude a few oddities such as "red beryl," which are so scarce as to have no real market at all. A flawless ruby, for instance, is worth more than a flawless diamond of equal weight. Because large rubies are so extraordinarily rare, their price jumps markedly with increase in carats. This is true for any gem, of course, but the price difference in ruby is really remarkable.

In 1646, the great gem-explorer Jean-Baptiste Tavernier, wrote tellingly of rubies: "At one carat, there is a price. At two carats the price doubles. At three carats the price triples." He concluded somberly, "At six carats, there is no price."

Today, the price for a single carat Burmese ruby can range up to $3,000 per carat, although flawed stones can go for considerably less. (The price was even higher before the 1992 discovery of the mine at Mong Hsu, deep in the eastern part of the Shan State, about 150 miles east and a little to the south of Mogok.) A two carat, high-quality stone can cost up to $8,000 *per carat*—that is, $16,000.

Great rubies come at great prices. For example, there is the Alan Caplan Ruby (sometimes called the Mogok Ruby), a 15.97 carat antique, untreated, cushion-cut Mogok stone. Caplan himself was an interesting figure. He was born in 1913 and orphaned at an early age, but he went on to become one of New York's most important gem dealers. He claimed to be Jewish, but was probably Irish (his real name was Callahan), who "became Jewish" because he felt the Irish were discriminated against in business.

Much more than a businessman, however, Caplan was a true geologist and rock hound, spending a youthful period of his life working at the Denver Museum of Natural History and excavating a mammoth along the South Platte River. He attended the University of Colorado (1934–37) and the Colorado School of Mines (1937–38). He also made about 10 trips to Brazil, scarfing up enormous quantities of topazes, quartz, and chrysoberyl. The war came (he served in Italy), but then he returned to Brazil and later the Orient for more buying. (Besides the Mogok Ruby, he at one time also owned the priceless Moghul Emerald.)

The Mogok Ruby was auctioned by Sotheby's for $3,630,00 on October 18, 1988, and sold to the London jeweler Laurence Graff. That comes to $227,301 per carat, setting a record price for a ruby at that time. Supposedly the stone was eventually purchased by the Sultan of Brunei as an engagement ring for one of his wives. Although its rather pinkish tone is lighter than is perhaps optimal, it has wonderful color saturation and is a truly valuable find.

However, the Caplan price record was broken in 1993 at the Myanmar Gems Enterprise mid-year auction in Rangoon. An unnamed gem, weighing in at 38.12 carats, sold for $5,860,000 ($153,725 per carat). This was the largest price ever paid for a single ruby up to that time, although it doesn't equal the carat-per-carat price of the Caplan.

But even that wasn't an end to the ruby-inflation. A *new* record price per carat for a ruby came to about $425,000 for an 8.62 carat cushion-cut, unheated Burmese ring by Bulgari, at Christie's "Important Jewels" auction. The exact source and date of its find are unknown. This totals to a stunning $3.6 million, paid again by Graff at the Swiss resort St. Moritz. Graff conceded the price was high but added that the cut and color distribution was the best he had ever seen. Graff decided to name the stone after himself, the Graff Ruby. Until this purchase, Graff had been more famous for yellow diamonds, but apparently he had caught ruby fever. On June 1, 2006, Graff bought at Christie's Hong Kong a 9.25 carat Burmese cushion cut ruby for $2,418,000 (about $260,000 per carat).

Star rubies have never sold for the prices of faceted gems, but good ones are not cheap, either. The auction record for a star ruby was for an unnamed Burmese 26.40 carat star ruby cabochon, sold at Christie's New York on April 12, 1994. It went for $1,080,500 ($40,928 per carat). The date it was found and its source are not known, nor are its present whereabouts. It is presumably not lost, but someone isn't telling.

Traditionally the value of gems is determined by the "Four C's": carat, cut, clarity, and color. However, only the first of these can be objectively stated.

CARAT WEIGHT

Gemstones, of course, are weighed in carats. A carat weighs 0.2 grams, or 0.007 ounces. Each carat is further divided into 100 "points." The word "carat" actually stems from the carob tree (the one from which Judas Iscariot supposedly hanged himself) that grows all around the Mediterranean. Its pods were once a major source of sugar for the region. Its connection with gem weight is that for centuries people used carob seeds as a standard for weighing precious stones. (For those areas short on carob trees, a carat was equal to four rice grains.) A carat equaled one carob seed. The original idea, apparently, was that the seeds of all carob trees were identical in weight, thus making them a perfect measuring tool. Alas, this turns out not to be the case. A research team from the University of Zurich, Switzerland, and colleagues from other research institutions collected and weighed 550 seeds from 28 carob trees in Mallorca and found that they are just as variable in weight as seeds from 63 other species of trees. This discovery seems to be only common sense, but now we know for certain. The carob tree is also a source for a substance its marketers claim tastes just like chocolate, but which does not.

Before the standardization of the carat, measurements were wildly erratic. The carat weight (200 milligrams) wasn't "standard" in the United States until April 1, 1914, and in the rest of the world not until 1930. This is the so-called "metric carat." One previous standard, the so-called international carat of 1877, was a little heavier, weighing 205 milligrams. Carat weight could even vary from city to city. For example, in one Italian city, Bologna, a carat equaled 188.6 milligrams; in another, Turin, it equaled as much as 213.5 milligrams. Across the Mediterranean, in Alexandria, Egypt, a carat was only 191.7 milligrams.

Consequently, older measures of how much a stone weighed might not be accurate by today's standards. Again, because density (specific gravity) is not the same for all minerals, stones having the same weight may well not have the same volume, just as a ton of feathers takes up more space in the kitchen than a ton of gold. This is why a one-carat diamond looks bigger than a one-carat ruby.

High quality rubies of over five carats are quite rare, and even rubies of two carats are very, very valuable. (It has been said that when the gods were dispensing rubies, they kept all the best ones for themselves.) The massive gems celebrated in medieval romances and in the literature of the Orient were either entirely fictional or were actually some other mineral like garnet or spinel. Another reason that large rubies are so rare is that even if they are found, they are almost invariably cut up into smaller specimens. The mineral market is not nearly as profitable as the gem market, and it's almost impossible the get the real "worth" of a large ruby crystal unless it is cut up into smaller gems. Indeed, intact large stones are usually of poor quality. One of the largest rubies ever found may be the 496 carat (originally 504.5) rough (about golf ball size) stone found in February 1990 in Dattaw, in the Mogok area. The massive crimson rose ruby was named the "SLORC" ruby. SLORC is an acronym for the "State Law and Order Restoration Committee," a euphemism (sort of) for the current dictatorship, which according to the Burmese Freedom and Democracy Act, "continues egregious human rights violations against Burmese citizens, uses rape as a weapon of intimidation and torture against women, and forcibly conscripts child-soldiers...in fighting against indigenous ethnic groups...including the Karen, Karenni, and Shan People." The said "Committee" took government control in 1988. At the time, the Committee made a big deal about how this stone was bigger than the Star of India. It isn't (the Star weighs 563 carats). Besides, the Star of India is a sapphire, not a ruby. The discovery of the SLORC ruby is a strange tale.

Apparently the owner of a brothel in Mandalay, Than Htun, was losing business, so he made an appointment with an astrologer to get to the bottom of things. The astrologer told him he was wasting his time in the prostitution business and that he should go to Mogok, where riches awaited him. Taking along a couple of associates, Than Htun made his journey, and lo and behold, he just up and discovered the massive (and supposedly flawless) stone. Than Htun immediately smuggled the stone to Bangkok, where he tried to sell the gem for two billion dollars (even a ruby that size would never fetch this amount). It wasn't long before Burmese agents learned about the stone, and simply arrested the families of the men involved. To protect their loved ones, the chastened miners returned to Burma, where they received a life sentence in prison, or were executed, depending upon whom you believe. At any rate, the stone was declared a national treasure. The crystal is now called the Nawata Ruby and is in the Rangoon Gems Museum. Sometimes people are even allowed to view it.

CUT

The art of "cutting" gems belongs to the lapidary, whose task is to bring out the beauty of the color and draw in the magic of the light. He is also charged, of course, with making the stone suitable for mounting in jewelry. Today, the word "cut" is actually a misnomer for most gems. While some stones, especially larger ones, are still sawn or cleaved, the standard treatment for most gems is simply grinding away at them with diamond dust or "grit." First, the stone is cut with a diamond laser or saw and then roughly shaped on an abrasive wheel. Then the faceting work is done. Some rubies are "cut" in cabochon rather than faceted. This is always done with star and cat's eye rubies, and with others for which it is decided faceting is not appropriate. Usually these are the more opaque, lower-quality stones. In ancient times, the mathematics and physics of perfect faceting were unknown, and so most stones were cut in cabochon to reveal their best color. In Europe, most stones were polished with sand, which has a hardness of 7 on the Mohs scale; however, some stones, including diamonds, rubies, sapphires, topaz, and spinel were too hard to really polish this way. It was not until the trade routes of the Roman Empire were established that "emery" appeared from Asia. Emery is simply granulated corundum. Thus, all gems except diamond were now susceptible to polishing.

After cutting comes faceting. Faceting requires great knowledge and an intimate acquaintance of the stone itself. The wrong type of faceting can greatly diminish the stone's beauty. The lapidary must be extremely precise at this phase. Finally, the stone is polished, a process that increases its luster, or reflective ability.

The final cut of rubies often depends on the shape of the rough stone. Rubies are first assessed for their symmetry by placing them face up, and cut in relation to that. The main parts of a modern cut for rubies include the crown or bezel (the upper parts), the pavilion (lower part), and the line that separates them, the girdle.

In India, stemming originally from magical reasons, the absence of flaws in a stone and its size are much more important that its sparkle. It is for this reason that although Indian gem cutting technology remained on a par with that of Europe, Indian lapidaries were loath to chop away at a ruby just in order to get more sparkle. They were interested only in cutting away the dangerous flaws in the stone, which detracted from its psychic power, but wished to leave the stone as large as possible. The bigger the stone, they believed, the greater its protective power. (This was true for any precious stone.) In fact, there is a sobering story about the Indian ruler, Aurangzeb, who had a 787.5 carat raw diamond. A Venetian lapidary, Hortenzio Borgio, made it shine, but in the process reduced its carat weight to 280. Aurangzeb fined him 10,000 rupees for "ruining" the stone by decreasing its protective powers. Although Aurangzeb was a famous gem connoisseur, Tavernier claimed that he had never seen the emperor actually wear any jewel other than a large topaz weighing 157.25 carats. Apparently the emperor was one for supreme understatement.

Unlike diamonds, in the world of colored gemstones there is no single perfect cut, geometrically configured for maximum brilliance. Cutting colored gemstones is a much more complicated proposition, in which the lapidary must consider the depth and evenness of color and ways to minimize inclusions. Unfortunately, many fine rubies have been utterly ruined by primitive and inferior cutting carried on in Burma. Naturally, the Burmese would like to maximize their profit by not only producing the rough, but by faceting as well; however, as of yet they don't really have the skill to carry this out properly. It should be noted that Third World lapidaries may charge under five dollars to facet a gemstone; it may easily cost 10 times that amount in Europe or the United States. Nowadays, it is becoming more common to calibrate the cutting of colored gemstones to even a tenth of a millimeter, as has been done with diamonds for years. This poses a problem for most "native-cut" stones, which are cut to maximize size at the expense of symmetry or precise sizing.

The most up-to-date lapidaries use computer-enhanced machines that help find the best cut for each stone. Some even offer a little grinding assistance. Cutting a gemstone is a delicate task, but it has been made much easier of late by computer-assisted three-dimensional imaging techniques. The lapidary first needs to orient the stone so that its finished cut will display the best color; he also needs to be on the lookout for ways to avoid flaws in the stone.

Nearly all gems are cut into some version of the step cut, the flat table-like cut suitable for emeralds, or the brilliant cut, which is best for diamonds. The step cut shows off color best, while the brilliant cut makes the most of the stone's fire. That puts ruby in an odd position, as it has both wonderful fire and glorious color. A compromise is called for. The best known cut for rubies is some version of the oval or cushion cut, which combines elements of the brilliant cut and the step cut. Typically, this cut has a rather low brilliant cut crown, divided into 33 facets, with a deep pavilion consisting of three tiers of stepped facets. Thus, while preserving many of the facets in the brilliant cut, the cushion cut has a more rounded "cushioned" appearance. This is also the "natural" shape of most rough ruby crystals, while the round or emerald shapes are rarer and may cost 10 or 20 percent more. Round shapes tend to enhance the brilliance of a stone, at least potentially, and are hence more desirable.

The marquise, oval, and pear-shaped cuts are merely variations on the brilliant cut. The marquise cut was named in honor of Madame de Pompadour (1721–1764)—it was suggested by King Louis XV as a reminder of the shape of her mouth.

Rubies and their sister-stones, sapphires, are also often cut with rounded profile below the girdle. For that reason these stones are sometimes called "bellied." Cutting bellied stones is one way to retain carat weight, but such a stone may be out of proportion, although no brilliance is lost.

On the other hand, pear and marquise shapes are less fashionable right now, and trade about for 10 or 20 percent less than ovals of the same quality.

Correct cutting not only maximizes the size of a finished gem; it also improves the "color coverage." Rubies cut too shallowly have reduced saturation in the shallow area (the so-called "windows.") Lapidaries sometimes purposely cut a slight window in the center of a dark ruby to enhance its liveliness against others when the wholesaler looks at an entire of parcel of stones. When the ruby is reset into a ring, the "window" is effectively closed and the stone darkens. Buyers looking at unset stones are often urged to hold the stone between their fingers, which imitates a setting and closes the window. This gives one a better idea of the true quality of the gem.

Overly steep cuts create dark spaces known as "extinction." These areas can be observed as the gem is "rocked" gently under a light source. Extinction occurs when the light enters the gem through the table, crown, or pavilion and then strikes an internal facet at such an angle that it does not return to the eye, resulting in "dead" or dark spots. This is why the proper cut is so essential. The balance is indeed delicate.

The "brilliant" areas of a stone are those in which the internal reflection allows the display of the most highly saturated colors. One of the most fascinating facts about Burmese rubies is that their strong fluorescence basically masks areas of extinction, a quality that adds to their value. Rutile silk, by scattering light to the facets, also helps mask extinction. Thai-Cambodian stones lack silk, thus showing more extinction.

The right cut increases the gem's sparkle or scintillation. Scintillation is largely created by the use of *small* facets. Generally, the larger the gem, the more facets it requires. A gem's "fire" or dispersion is created when white light separates into its spectral colors as it passes through two non-parallel surfaces (such as a prism). It is largely dependent on the composition of the mineral itself. Even stones with a naturally high dispersion, if they are color saturated, can "mask" the fire. Corundum has a low dispersion factor, however, and the rich color of the best rubies makes dispersion an unimportant factor to consider when cutting or evaluating this gem.

COLOR

In most cases, the redder the ruby, the more it is worth. And that is the simple truth.

Still, not all ruby experts agree as to the desirability of a pure red for rubies. Some think that a slight secondary hue of orange, pink, blue, or violet can add a certain interest and sparkle to the gem. They also disagree about what the best secondary hue should be. It's simply a matter of preference and of what is fashionable at the moment.

A lot of this depends upon cultural values. For example, it is said that Australians like a darker ruby than do some other people, as long as the cut and clarity are good. The United Kingdom is mostly a market for smaller, darker stones of poorer quality. (This predilection is also reflected in the British taste for garnets.) The French seem to be pickier, preferring the more classic, rich, fire-hued Burmese stones. Nothing is more important than color to the French. Germans, on the other hand, seem to place a premium on a good cut and clarity, with less emphasis on color. When color is considered, Germans seem to like a pinker stone than do the French. Unlike their neighbors, the frugal Swiss aren't much for buying rubies for themselves (although Geneva is a major center for ruby-buying and selling to rich foreigners), so it is hard to gauge what they might like. The Japanese have similar tastes to the Germans, and are not usually interested in darker stones at all. Clarity and cut are both important to them. Italians prefer a classic rich red stone, and will sacrifice clarity to get it. They are also more willing to give up excellence in cut for a shallower stone that has a bigger "face." In the Middle East, most buyers are interested in big stones and gaudy jewelry. Quality takes a back seat. As for the huge American market, almost anything goes, although the most popular stones are a good, balanced blend of color, cut, and clarity.

In color as in everything else, the Burmese ruby is and has always been the "gold standard" for rubies. They are the only stones that most frequently possess all the qualities one expects from a great stone: pure, saturated color, fluorescence, and few inclusions. Thai stones have the color, but not the fluorescence. Sri Lankan stones have the fluorescence but not the color; African and Afghan stones are seldom "clean" enough to facet. Whatever the desired shade, a good ruby holds its color well in all lights.

When they are being sold in their native lands, rubies are traditionally shown on brass plates or yellow backgrounds, which cancel out the blue notes present in many rubies and make them appear more red. This is an old ruby-sellers' trick. To get a better look at the real color, rubies should be shown against a white background in incandescent light or natural skylight (not direct sunlight). Fluorescent light is death to rubies, since fluorescent tubes emit no red and severely diminish the red appearance of the stone.

Rubies are responsive to weather and light conditions. The quality of the light has a slight, and to the expert, pronounced effect upon the color of a stone. To really see a stone, an expert wants to look at it by the uncompromising natural light at midday, especially under the north light (in the northern hemisphere). Rubies also tend to look their best in midday and under cloudy conditions, and so should be examined at different times. (Sapphires, on the other hand, should be looked at in the early morning or late afternoon under blue skies. This is not a matter of legend, but of optical conditions.) Light colored rubies look best in dim light, while darker ones explode with fire in the brilliant light of day. There are, of course, special optical lamps to simulate desired conditions, such as short-arc xenon lamps, but they are not cheap. Ruby's imitator, the spinel, may have the same color as ruby, but is not pleochroic, doesn't have the rutile silk effect, and looks less bright in strong light. Another imitator, garnet, is not pleochroic either, although its luster or shine can be very ruby-like. (Since garnets and spinels are singly refractive, meaning that they do not split light, they cannot be pleochroistic.)

Even latitude makes a difference. The strong sunlight of the tropics, where the majority of rubies originate, shows up a darker looking stone than more temperate light. Then there's the Purkinje shift, a phenomenon that makes the eye more sensitive to red in bright light.

CLARITY

Clarity refers to the transparency and lack of inclusions in the stone. It is a virtue in any gemstone "meant" to be transparent or translucent. As a general rule, the clearer the stone, the more value it has. Opaque or semi-opaque rubies are not worth very much, even if they show asterism. Buyers are often told that when the choice is between a stone of good color plus inclusions and a clear stone of poorer color, to "go for the color."

SETTING THE STONE

After a stone is cut, it is set. And although Paracelsus (1493–1541) declared that rubies had an "affinity" to silver, most contemporary jewelers and their customers agree that the natural metal for ruby is gold.

While setting a single ruby can be a routine task requiring no more than mediocre skill, producing a matched set is on a different order altogether. Unlike the colorless diamond, matching the infinitely shaded ruby is an extremely difficult job, especially when you consider that size and shape must also be matched; it is a process that can take many months, especially for the higher quality stones.

The ancients maintained that the best setting for a ruby was a mixture of copper and gold, but that the stone should be set so that it actually touched the skin of the wearer in order to impart its magic.

5

The Mines of Myanmar and Elsewhere

Rare they may be, but rubies pop up, at least occasionally, all over the planet. However, the best quality and greatest numbers come from Burma (now officially Myanmar, except to the U.S. State Department), which produces a beautiful, clear, deep red stone. It has done so for almost 1,000 years.

In fact, about five-sixths of the world's rubies originate there. For ruby aficionados everywhere, the word "ruby" is intimately associated with Mogok, a town in the Katha district of northwestern Burma. The rubies of Mogok are famed for their wonderful fluorescence and the perfect amount of rutile silk that makes them glow with inner light and fire. One legend claims that rubies can never be covered by cloth, as their inner fire will burn through any material. Thus, the ruby can never be concealed, a fact which must have amazed the smugglers who sneaked through with them on a regular basis.

Burma is a Buddhist country, and Buddhism discourages the use of jewels for personal adornment, although Burmese monarchs were dripping with gems, despite the religious injunctions against the practice. It was also perfectly acceptable to cram precious stones into the *htarpanar-taik*, or relic chambers of pagodas and other Buddhist shrines. Many great and glorious gems were preserved in this way. One of the most famous is the cone-shaped Shwe Dagon Pagoda, probably built about 180 BCE by Mon Buddhists. The pagoda itself is a solid brick stupa completely covered with gold. Rudyard Kipling called it a "golden mystery...a beautiful winking wonder that blazed in the sun" (*From Sea to Sea and Other Sketches: Letters of Travel*, 1899). The sixteenth century English traveler, Ralph Fitch, in his "Account of Pegu in 1586–1587," remarked that is of a "wonderfull bignesse, and all gilded from the foot to the toppe. . . It is the fairest place, as I suppose, that is in the world."

It is said to contain relics of the past four Buddhas enshrined within: the staff of Kakusandha, the water filter of Konagamana, a robe fragment from Kassapa and eight strands of the hair of Gautama, the historical Buddha. The hair bit is somewhat of a puzzle as Buddhist monks are known to have shaved their heads, but perhaps these were old hairs from the days before the Buddha became enlightened.

The relic chamber and the *hti* (a special chamber built for such purposes) contained many rubies and lots of gold. King Mindon Min added many more (as well as a 70 carat diamond) during his tenure in the nineteenth century. One legend tells that eight Buddhist holy men went in search of a ruby reputed to have mystical powers, and during their search, stopped for a rest at the place where the pagoda now stands. Later they found the ruby (and another sacred hair of the Buddha in Syria) and came back to build the pagoda on the place where they had rested. The ruby and hair went into the relic chamber.

Besides Mogok, sources for Burmese rubies include Sagyin (near Mandalay in the center of the country), Thabeitkyin, Naniazeik (near Myitkyina), and most recently, Möng Hsu (located in the central Shan Plateau).

Burma is a medium sized place, as countries go, slightly smaller than Texas, bordered by India, China, Laos, and Thailand. Its most well-known geographical feature is the Irrawaddy River, which bisects the country from north to south. The gem-bearing part of the country, the so-called Stone Tract, in the famous Mogok Valley, is in the central northeast. Before World War II, the Mogok Valley was divided into two areas—the Mogok Plain and the Lower Mogok Plain. The latter area was said to produce the most excellent rubies, with invisible rutile and special brilliance. Stones from the Upper Valley were darker, and produced an undesirable bricklike shade when cut. The original word, Mong Kut means "winding valley" to the Shan people who named it. "Mogok" is apparently a Burmese corruption of the name.

In the seventeenth century, Tavernier wrote of Burma, "It is one of the poorest countries of the world; nothing comes from it but rubies, and even they are not so abundant as is generally believed." He adds, "All the other colored stones in this country are called by the name ruby and are only distinguished by color" (*Travels in India*, Book II, chapter 19). Today, even the Burmese tourist trade, what there is of it (and which at times is officially non-existent), is linked directly to its wealth in red stones. However, it has so much natural beauty, diverse culture, and fascinating wildlife (including elephants, tigers, bears, and leopards) that eco-tourism is bound to play an increasing role in its economy (once the political situation in Burma stabilizes). It has always been rather isolated. In all his six journeys to Asia, Tavernier never went to Burma, the ultimate ruby source.

The entire area is nicknamed "the valley of rubies," but most famous section is the Mogok Stone Tract (Yetaana Twet Mye). This L-shaped area lies in the northwest end of the Shan Plateau. The vertical leg of the L as you might expect, reaches south to north; the horizontal arm follows the famous Irrawaddy River. The mining district covers about 1916 square miles in all, although only about 70 square miles of that is thought to be gem-bearing. The valley is drained by the Shweli River, which empties into the Irrawaddy.)

Legend says that the "modern" city of Mogok was founded in 579 CE by a tribe of headhunters from nearby Momeik (Möng Mit). They got lost, and in their efforts to find their way found a mountain gorge filled with rubies (and snakes). Since the valley was guarded by deadly serpents (its frightening nickname even today is "valley of the serpents") and was too full of pestilent marshes to enter, the clever Burmese threw lumps of meat (not human, one hopes) down into the gorge. The smell of the meat drew birds of prey who grabbed the meat, along with the rubies that stuck to it.

These stones were then transported to the birds' nests, where they were easily retrieved by the hunters. There are some elements of the tale that may not be strictly true, but it makes for interesting reading. In fact, much the same story is told elsewhere, including a version in *Sinbad the Sailor*, compiled about 800 CE ("The Valley of Precious Stones"). The ultimate source may hark back to a story told about Alexander the Great, but in that case the birds were vultures rather than eagles. All these stories were explained by Dr. Valentine Ball (Tavernier's biographer and early editor) has having originated in the ancient custom of sacrificing cattle when new mines were opened, and leaving some of the meat on the spot as an offering to the guardian gods. When the vultures flew away with the meat, the story got around that they were also taking away some rubies that were attached to them.

While the traditional birth-date of the village goes back only to 1207, pre-historic (Middle Pleistocene era) tools made of jadeite found in the mining areas of Burma indicate that our passion for this gem goes back many thousands of years, although of course we have no trade records from the Stone Age.

However mythically it was founded, Mogok's location is factually known. It lies 644 kilometers (about 400 miles) north of Rangoon, the capital, and 210 kilometers northwest of Mandalay. It lies about 1500 meters above sea level, in a secluded mountain valley surrounded by heavily jungled, orchid-blooming hills that rise to 2347 meters and surround the town like an amphitheater. Its population is probably between 300,000 and 500,000 of various ethnic groups, including Burmese, Shan, Nepalese, and Lisu people, with a variety of faiths—Buddhists, Hindus, Christians, animists, and Muslims. About 90 percent of them still earn their living in the gem business. Some of it is legal and some is not.

Gems have always been critical to the Burmese economy, although due to smuggling and intermittent bans on export, it is impossible to estimate just how much money they do generate. It is also a fact that the Burmese fail to make the most of this precious natural resource. Despite the vast number of rubies and other gemstones the country has produced over millennia, the country has no sophisticated cutting or polishing facilities.

The famed Mogok area contains precious stones other than rubies, too, although one might never know it, for all the attention paid the premier stone. An anonymous traveler, writing in 1905, commented, "No matter what business may have brought you to Mogok, the natives all assume you are there for rubies—rubies, nothing but rubies" ("A City Built on Rubies"). Spinels of all colors, peridot (the "Mogok emerald"), moonstone, lapis lazuli, zircon, peridot, aquamarine, and even, more rarely, sapphires can also be found. But the ruby is king of them all. (At any one time it is believed that between 700 and 1500 mining leases are issued in Mogok—depending upon the political situation. In calmer times, more licenses are issued.) Mogok is indeed, a city "built on rubies," for it has no other real industry.

Mogok belongs to the Shan area, the largest state in Burma. It borders southern China, western Laos, and northern Thailand. It takes its name from the Shan people, who are the ethnic majority. First annexed by the king of Pagan, Anawrahta (ruled 1044–1071), Shan has, at divers times, been independent of Burma and a part of it. Anawrahta also established a diplomatic relationship with Sri Lanka, collecting important religious artifacts from that land in exchange for fine Burmese rubies. So it is that some "Sri Lankan" rubies are originally Burmese. The chieftain (saopha) of Mon, Saw Naung, was anxious to develop a good relationship with Anawrahta. To this end, he offered up his sister Saw Mon Hla as a bride for the king. The event is recorded in song and story to this day, and rubies play a central role. She arrived encrusted with jewels, silk, and gold cloth, but her earrings were what attracted the most attention. They were characteristic of the Mon style—and were properly called "ear-tubes," which were made of metal and covered on three sides, with the open end facing front and carrying stones at least 3 centimeters by 2 centimeters. (A pair of such ear-tubes are on display at the Victoria and Albert Museum.) The king was duly impressed and commented on the shining lights of the ear-tubes, most likely a result of the daylight fluorescence of the stones.

Enemies of Saw Mon Hla declared that the strange lights were of ill-omen and represented her dark magical powers. Perhaps she was even a sorceress and should be asked to leave Pagan immediately. The superstitious king heeded this advice and sent the lady home forthwith. On her way back, it said that she stopped at a river to bathe, removing the troublesome ear-tubes to do so. When she came back to retrieve them, she found that they had "moved" and were gleaming with a very strange light. (Since they had been in her ears until then, perhaps she had just never noticed it before.) Terrified, the princess vowed that if allowed a safe return home, she would have a pagoda built there, and include the ear-tubes as part of its treasure in the relic chamber. (She was probably glad to be rid of them.) All this was accomplished and the Shwe Zaryna Pagoda still exists, near the present day city of Mamyo, in northeast Burma.

Anawrahta's dynasty came crashing to a halt (during the reign of Narathihapate) when Kubla Khan invaded Burma in 1271. He sacked Pagan (a depressingly frequent occurrence to its inhabitants) and looted the relic chambers of the beautiful pagodas. At the height of its glory, Pagan had 6,000 pagodas, of these about a thousand remain today. The city is now considered a World Archaeological site. Rubies were so prized by the inhabitants that whenever a large one was found, the entire court got the day off.

Anawrahta also conquered the Mon Kingdom of Thaton, and it is said he brought back to Pagan an immense, oblong-shaped star ruby. Manhua, the king was allowed (for a time) to keep some of his possessions and prestige. At one point Manhua decided that he should build a pagoda, and passed a tray around to his family asking if they would like to contribute anything to the construction. At that time (somehow) Anawrahta, turned up and slipped onto the tray a fabulous star ruby set in a peculiar setting. Now where had Manhua seen that ruby before? Correct, in his own palace. At any rate the ruby was sold for the making of the pagoda and has never been seen since.

Then Anawrahta left the country, apparently finding nothing more of interest. For two hundred years, Burma remained without strong legitimate leadership, and the entire country was overrun with dacoits (bandits) who ripped off (literally) every precious gem they could find, breaking into relic chambers and gouging out the ruby eyes of the animal statues set up to guard the pagodas.

Currently the Shan State is once again part of Burma, having been attached following World War II. Between 1287 and 1555, the Shan people ruled Burma, first from Pagan and then from Ava, which would remain the Burmese capital for about 500 years. However, centuries before this, Shan was a client state of Burma. It is reported that in the sixth century CE, for instance, the ruler or *saopha* of Momeik in Shan was accustomed to sending out about seven pounds of rubies to the central government each year as tax-tribute.

To get out of paying tribute (a practice apparently going on for centuries), Momeik attempted to strike a deal with the king of neighboring Yunnan. (This king was sometimes viciously known as the Frontier Eunuch. In return for protection, Yunnan would get a share of the ruby mine profits. However, the weak saopha who made the agreement died, leaving his daughter-in-law Nang-Han-Lung in charge. She was in no mood to pay off the Frontier Eunuch, although she did send him a nice ruby as a present. She sternly went about getting the country into shape between 1450 and 1484, reorganizing the 13 villages of the Mogok Stone Tract into a single community. She also declared that all the rubies belonged to her. But she also sent them to other rulers and pretty soon everyone was at war. Eventually, in 1484 the Stone Tract was ceded, although not formally annexed to the Burmese Kings. The formal annexation occurred later, during the Second Burmese Dynasty.

Later, according to the Royal Edict of 1597 (the first known reference to ruby mines) the Burmese king, Nuha-Thura Maha Dhama-Yaza, forced the *saopha* of Momeik and the Shan states to trade Mogok and Kyatpyin (probably the legendary "Caplan") for Tagaungmyo, a trade which did not favor the prince, to say the least. However, apparently the luckless prince was unaware of the treasure in his own backyard. Nuha-Thura Maha Dhama-Yaza simply annexed the entire area. Thus the valley of Mogok became part of the Golden Land, as Burma was sometimes called. And with good reason. The gold belt is located in the northeast, extending eastward to the headwaters of the Irrawaddy River, north to the foothills, and south and west of the Hukaung Valley and Bay of Bengal. There are also deposits of gold in the state of Momeik, in Pegu region and elsewhere. It is this gold that was used to gild pagodas and palaces all over the country. Even today, from a quick look at a map of the area, it can be seen that the Stone Tract Area was simply gobbled up, and doesn't seem naturally a part of the rest of the territory.

This dynasty, the second or so-called Toungoo dynasty (1541–1661), got its start with Tabinshwehti, who conquered the Shans in 1541, then took over most of the rest of the country. He even went after the Portuguese at Dalla and Martaban, beat them, and set up his own capital at Pagan or Bagan (not to be confused with Pegu). He was succeeded by his general (and brother-in-law) Bayinnaung (1551–1581), who further expanded the kingdom. His main object of course was the tiny but rich state of Momeit, whose saopha fled into the jungle to escape the invaders, to no avail. Starving and desperate, he finally surrendered to Bayinnaung in 1562. He was chained up and sold as a slave to the Maharaja of Ganges. Bayinnaung was not finished, however. He then attacked Thailand, from which he brought home skilled jewelry artisans. (One of his objects in attacking Thailand had been his attempt to snag for himself the king's Royal White Elephant, a symbol of good fortune.) He also attacked Laos and Yunnan (of Frontier Eunuch fame). Bayinnaung removed the capital to Pegu.

As is evident, Bayinnaung was a vicious man with few scruples. He even took rubies from the casket of the former and deeply beloved Mon monk-king of Pegu, Dhammazedi (reigned 1472–1492). Perhaps from guilt, he donated several rubies to the Shwe Mawdaw Pagoda there—where they remain to this day.

Bayinnaung had his romantic side, of course. The story goes that he fell deeply in love with Hne Ain Taing, a rich young Mon widow. She already had one son, but became pregnant again. At the news, Bayinnaung suddenly found out he was wanted on emergency business at the capital and had to leave, but gave the woman a ring and told her to stop by his palace some time if she bore a son. (If it was a girl, he added, Hne Ain Taing might just as well stay home.) He did have a very nice palace, well worth visiting. It was as big as a city block and made of the finest teakwood. Its columns were encrusted with jewels, and the moat around it was well stocked with crocodiles. He got part of the money to build it from the saopha of Momeik, an offering that, as we have seen, did him no real good.

Hne Ain Taing bore not one but two sons. Later on she showed up at the palace with both of them as well as the ring, but the king was still hazy about the whole affair and made her go through a water ordeal, undoubtedly involving crocodiles. She passed and he acknowledged the younger child as his own and named him Shin Thits.

This whole story is reminiscent of the Sanskrit drama *Shakuntala* and other traditional tales of the day, and so is probably fiction, although it's possible the king modeled his behavior on some literary character. At any rate, despite Bayinnaung's disgraceful behavior on several counts, he is still regarded as a great hero in Burma. Even Sir James George Scott, in his *Burma from the Earliest Times to the Present Day* (the "present-day" in this case being 1924) called his reign "the greatest explosion of human energy ever seen in Burma." Scott (1851–1935), was an interesting character in his own right. An acknowledged expert on Burma, he attempted to increase his credibility by pretending he was Burmese and writing some of his books under the pseudonym Shway Yoe.

After the death of Bayinnaung, the country was rattled by war. Finally, it was taken over by Myo-Thugyi (also known as Alon Prom), who founded the third and last Burmese Dynasty in 1752; it lasted until 1885. The new king returned the capital to Ava and went on to conduct a series of foreign wars in the time-honored fashion of the earlier Burmese kings. His descendants followed suit.

Towards the end of the second dynasty emerged the story of the Nga Mauk ruby, which hailed from Mogok. This stone is considered the oldest and best known piece of the ancient Royal Burmese collection. It was found during the reign of King Pindale (1648–1661).

According to the usual version of the story, the finder of the original 560-carat stone was a poor villager named Nga Mauk, after whom the stone was named. He was just wandering along the riverbank, and there it was, so brilliant that it caught his eye even though it was on the other side of the river. Nga Mauk immediately swam across, picked up the stone, washed it off, and examined it in the light of the sun. His whole hand turned blood red with the inner fire of it. Nga Mauk hid the stone in his longyi, the ubiquitous one-piece garment of Burmese villagers, and swam back with it. By most accounts he then presented it to the village headman, who suggested they take it immediately to the royal treasury. (Another story claims that Nga Mauk's wife traded the stone for a rupee's worth of fish condiments, but it's an unlikely tale and doesn't explain how Nga Mauk got the stone back.)

In any case, the headman and Nga Mauk presented themselves to the treasurer, who decided it should be shown to King Pindale then and there. The king was so shocked by the size and brilliance of the stone that he tried the classic Burmese test of proving a ruby's value—he put it in a glass of milk. Indeed, as predicted, the milk turned blood red. To authenticate it further, they placed the ruby on a cloth in the sunshine, where it glowed like a supernatural lamp. The real cause of the glow, of course, would have been the natural fluorescence characteristic of Mogok rubies. However, spinels can fluorescence in natural light as well, so it wouldn't have been a definitive test separating the two gems. Of course, it's not clear that the ancient Burmese even cared which was which. Mock-rubies like spinels have been featured in the crown jewels of England and Iran for centuries.

Another story tells a slightly different tale. It claims that the Nga Mauk ruby was found cracked almost in half. Nga Mau took advantage of nature's offering, as it were, dutifully sending one half to the king. This stone later became the chief stone of his collection, and was frequently seen sparkling in the ear of his favorite elephant.

Secretly, however, Nga Mauk sent the other half to Calcutta for resale. The king soon found out about what he regarded as treason. He made his displeasure known by ordering that all the inhabitants (including women, children, and Buddhist priests) of Nga Mauk's village be crammed into a stable and burned alive, with the inhabitants of neighboring villages forced to watch. This was something of a tradition among the Burmese, apparently. What was left over from this massive cremation can be seen at a place called Laugzin, which can be translated into the "fiery platform." Nga Mauk's wife, Daw Nann, escaped the inferno, watching the horrors from a distance and perhaps clutching her fish condiments. Today this hill is known as Daw Nann Gyi Taung ("the hill where Daw Nann looked down"). Today, standing atop this hill, one can see the Inn Gaung and Kyauk Phyu mines. You can also see the tin roofs of the villagers, a sign of prosperity.

The second half of the original stone was purchased (at an extraordinary price) and returned to Burma, where it "fit" the other half perfectly. Both pieces were cut in Mandalay. The first, as we have seen, is the Nga Mauk ruby (98 carats), and the second, weighing 74 carats, the Kallahpyan, meaning "returned from India." The carat weight is disputed, however. Whatever its weight, the second stone was cursed, as its was believed that crossing the black (Kallah) waters of the ocean resulted in caste reduction. Cursed or not, single or double, the ruby or rubies were added to the king's collection—and a new myth arose: that whoever possessed it would be granted supernatural powers.

Almost the same story is told of the 400-carat Maung Lin rough ruby, also supposedly found in during the reign of Mindon Min (1853–1878). The same tales are often told about two different historically famous stones. Or perhaps they are the same stones hiding behind different names. Since the stones have often vanished from history, it is impossible to know. Maung Lin is the name of the trader who handled its transfer. He bought it for 3,000 rupees or about 200 pounds. The gem itself was found by men working on the road to Momeit. This stone, however, was broken into three pieces. Two of these were then cut into stones of 70 carats (sold in England) and 45 carats (sold in Mandalay). The third piece (weight unknown) was sold secretly and uncut in Calcutta for 4,666 pounds. It was re-purchased at a huge price and returned to the land of its origin. Actually, it is even possible that the Maung Lin and Nga Mauk stones were the same stone or that some similar mix-up occurred. The location of the Maung stones is, of course, also unknown at the present time.

MINING THE STONE

Despite the romantic stories, few great rubies are found just lying on a riverbank. They must be dug up.

The simplest method of mining rubies is millennia old—panning for them in streams and old gravel beds that used to be streams. This is an excellent way to go looking for any heavy mineral like rubies, sapphires, and of course, gold. In Burma, the mining folk most often depend upon the monsoon (which supplies the needed water) for the mining season to begin.

In addition to simply panning for stones, various pit methods (used mostly in the valleys) have been tried, including small round *twin-lon* pits or larger pits called variously *lebin* or *kobin*, which are more common.

Since the valley alluvials are no longer as profitable as they once were (having largely been worked out), open-trench mining or *hmyaw-dwin* ("hmyaw") is used for hillside deposits or where the rubies are near the surface. The *lu-dwin* ("lu") is a method whereby gem-bearing materials are extracted from limestone caves and fissures; these have produced some of the richest finds, but this method is easily the most dangerous kind of mining. One such cave was so large that *hmyaw-dwin* and *twin-lon* were set up *within* it. Then the roof caved in, a not-infrequent occurrence. Even when the roof doesn't cave in, climbing down into deep cave cracks has resulted in numerous accidents. There is also tunneling directly into rocks, which has been done since the British first moved in.

In most cases, the stones are located in narrow crevices in the bed rock, and a pick or spade must be used to move the larger rocks. Luckier panners have hand-held pumps they can use in narrow spaces. The panner simply scoops up some likely looking material in a bucket and shakes it gently into sieves or wire pans. Often two sieves are used, a coarse sieve with holes of about 5 mm atop a fine sieve with holes of about 2.5 mm. The pan is placed underneath. The sample material is heaped into the top sieve, and both sieves and the pan are worked in a circular motion in the water. This will concentrate the heavy material towards the center and the lighter material around the edges. This way, when the pan is flipped, the heavy gems will more likely be found at the top. The gemstones can often be picked out with tweezers.

This kind of mining is possible only with durable gems not likely to shatter.

Because gemstones are heavier than most other minerals, they tend to settle at the bottom of the pan, while lighter material washes over the lip. For larger operations, motorized equipment such as dredges or suction pumps can be used to draw the gravels ("byon") into a hand-operated pipe that discharges them over "riffle boxes" or "jigs" that separate the material by gravity.

After extraction, the gem-bearing material is washed to separate the rubies from the lighter material. The government owns a central washing plant, but most of the larger mining operations have their own. Since time immemorial, the tailings, or mine waste, have been allowed to be picked over by women (and only women), who hoped to find a missed ruby. Any male who even stooped down to pick one up was subject to imprisonment. This merely resulted in the oppressed people of Burma working together. A worker spots a likely stone, gives a nearby woman a hidden signal and an indication of where it is, and she immediately "finds" it, later splitting the take with the worker. The custom is inviolate, however, even though the British, among others, attempted to forestall the stealing of stones by the workers themselves by forcing them to wear steel masks so that the gems could not be swallowed.

The miners got to keep the dust and the worst, smallest stones. Of course, this led to the practice of the miners' smashing up any large, good stones they found so that they could keep at least some of their findings. The British believed that only civilized people like themselves had a right to the gems of "the savages" and that they were permitted every brutality in the book to get them. The torture and death that ensued for evading taxes or smuggling had forced most of the population to flee. Finally, conditions got so bad that the king decided that the best thing to do was to lease the mines to the French. Then they could deal with the problems.

At least in legend, the finding of rubies is also closely associated with caves and their mythological inhabitants—dragons. More than one notable Mogok ruby has had some variety of the "Naga" or "serpent-dragon" incorporated into its name. The Nagas were supposedly an entire race of serpents. They were the children of Kadru, chief wife of the sage Kasyapa. They dwell in the nether regions of the earth, where they keep Vinata, Kasyapa's other wife, a prisoner. The great majority of Nagas are thought to be demonic, and they wear brilliant jewels in their hoods, which are so bright they light up the entire under world. In fact, it is said that they own the best jewels in all three worlds—upper, middle, and lower. Not even the holy gods can boast of such fine stones.

The myth hides a well-known truth. It is a fact that rubies originally occur in granular limestone that covers the Mogok hills, forming caves. But since limestone is soft and crumbly, these outcrops have been continually eroded and weathered throughout the millennia, and many gems have been washed down in alluvial deposits in river beds and basins.

The ruby-bearing caves are sometimes mined as well. Indeed, during the course of mining, fossils of elephant bones or teeth are often found. The early miners called these the Hsin Te Kyaik—"the vault of giant dead elephants." The people sometimes refer to these as "nagá ajó" or dragon bones, of which there appear to be several varieties. In one story, cave fossils of large animals were so frightening that the villagers shut the cave up for fear the deceased's relatives might come charging out of the cave to plunder and destroy the hamlet. Objects not designated as dragon bones are sometimes considered to be one of the Buddha's teeth and transported to a shrine for worship, although the monks disapprove of such superstition. Chinese culture also considered fossils to be "dragon bones," but the practical Chinese, instead of setting them up for worship, ground them up for medicine.

No matter whether they are mined in a conventional way or stolen from dragons, most rubies end up on the gem market. Therefore, another critical part of the Burmese ruby trade is the brokers. Every dealer needs such people, not only to help evaluate the stones but for information purposes. Brokers know who is buying what and at what price. There's a large psychological component to all this. Many deals are made in the dead of night, in snake-infested jungles. With the only illumination coming from a smoking torch or shaky flashlight, the gleaming stones are traded. With visibility conditions so imperfect, one wonders who is getting what, but it's an old custom. Even perfectly legal stones are traded this way. Some brokers go to great length to assure their clients that the rubies are in fact stolen and that the deal must be concluded secretly. The buyer apparently thinks he is getting a deal this way. After all these shenanigans, the gem appears on the market

THE FIRST EUROPEANS AND THE BURMESE RUBIES

The fifteenth and sixteenth centuries were the Age of Exploration, and Burma got its share of adventurers, missionaries, merchants, and soldiers (and soldiers of fortune). The overpowering beauty of the land, strangeness of its customs, and reputed wealth of its royalty were a siren call that few true-hearted explorers could ignore. Whether the goal was to buy or baptize, explore, extract, or simply experience—the magic of the land and its people was enthralling.

The first European traders to Burma (about 1400 CE) were interested in spices, which were literally worth their weight in gold, not only as taste enhancers but as food preservatives. However, when reports of rubies started filtering in from travelers like Nicola di Conti, Ludovico di Varthema, Hieronimo de Santo Stephano, and Caesar Fredericke, interest shifted to jewels.

Nicola di Conti (1420–1444), was the first European to visit Pegu, an ancient fishing port a little north of Rangoon in 1435. Although Pegu's king Binnyaran (1426–1446) did not actually permit Di Conte to travel to Ava, the old capital, Di Conti wrote about it anyway. "The King [Pyusawati] rideth upon a white Elephant, which hath a chain of gold about his neck, being long unto his feet, set full of many precious stones," he gushed (reported by Richard Hakluyt, 1506). Fifteenth century European travelers reported that the King's "idols" (presumably statues of the Buddha) were "covered" in rubies.

While the royal jewels of Burmese kings were famous throughout Asia, it is really impossible to know very much about the value of the Burmese royal rubies. In the first place, the person of the king was considered so holy that he and his clothes could be viewed only at a distance. It would not be possible even for an accomplished lapidary to tell much at a distance of several yards or even feet. It has been averred that a great many of these royal stones were (a) of poor quality (b) glass, or (c) imaginary. On the other hand, they could have been very genuine and very valuable. If anyone would have owned and worn such stones, it surely would have been the king. It's simply impossible to tell at this distance—and time.

"In Pegu," wrote the early sixteenth century visitor Duarte Barbosa, "they know how to clean but not how to polish them, and they therefore convey them to other countries, especially to Paleacate, Narsinga, Calicut and the whole of Malabar, where there are excellent craftsmen who cut and mount them" ("The Route Book of Duarte Barbosa").

Both Di Conti and fellow explorer Ralph Fitch apparently thought that rubies were actually mined in Pegu rather than Mogok; at any rate, they were not allowed to visit the actual mines, for understandable reasons.

Hieronimo de Santo Stephano arrived in Pegu in 1496, Like everyone else, Stephano was not allowed to visit the mines.

Di Varthema, who hailed from Bologna, Italy, reported that in return for a gift of coral, he received from the king of Pegu about 200 rubies. Coral, like rubies, was prized for its magical, medicinal power, as well as ornamental use. Like the others, he saw and described the trade in rubies, mentioning the source as "Caplan" or "Capellen" (Kyatpyin), 30 days' travel away. "Not that I have seen it," he confessed frankly, but added that he had heard it "from merchants." "The sole merchandise of these people is jewels, that is, rubies," he confided.

Alexander Hamilton, not the first American Secretary of the Treasury, but an earlier writer of the same name who traveled to the East Indies between 1688 and 1723, disagreed that the *only* merchandise of Burma was its jewels.

The product of the country is timber for building, elephants, elephants' teeth, bees-wax, sticklack [shellac], iron, tin, oil of the earth, wood-oil, rubies the best in the world, diamonds, but they are small, and are only found in the craws of poultry and pheasants, and one family has only the indulgence to sell them, and none dare open the ground to dig for them...the Armenians have got the monopoly of the rubies. ("A New Account of the East Indies: Being the Observations and Remarks of Capt. Alexander Hamilton," C. Hitch and A. Millar, 1744)

He also has some admiring remarks to make about the costume of the female inhabitants of that land:

Under the Frock they have a Scarf or "Lungee" doubled fourfold, made fast about their Middle, which reaches almost to the Ancle, so contrived, that at every Step they make, as they walk, it opens before, and shews the right Leg and Part of the Thigh.

This costume, he avers, was invented by a former queen of the country, who was attempting to cure the wicked men of the land of their attachment to sodomy. (Hamilton also wrote a fascinating account of a favorite Indian method of execution—crushing by elephant.)

Equally revealing (in another way, of course) are Di Varthema's comments about the King of Pegu.

...[H]e is so humane and domestic that an infant might speak to him, and he wears more rubies on him than the value of a very large city, and he wears them on all his toes. And on his legs he wears certain great rings of gold, all full of the most beautiful rubies; also his arms and his fingers all full. His ears hang down half a palm, through the great weight of the many jewels he wears there, so that seeing the person of the king by a light at night, he shines so much that he appears to be a sun.

Nor was the King of Pegu the only bejeweled king abiding in this land of jewels.

Caesar Fredericke had his own (much broader) experience, as he grandly informed his intended reader—in a sentence of true epic proportions.

Having for the space of eighteen years continually coasted and travelled over almost all the East Indies, and many other countries beyond the Indies, both with good and bad success; and having seen and learned many things worthy of notice, which have never been before communicated to the world; I have thought it right, since the Almighty hath graciously been pleased to return me to my native country, the noble city of Venice, to write and publish this account of the perils I have encountered during my long and arduous peregrinations by sea and land, together with the many wonderful things I have seen in the Indies; the mighty princes that govern these countries; the religion or faith in which they live; their rites and customs; the various successes I experienced; and which of these countries abound in drugs and jewels: All of which may be profitable to such as desire to make a similar voyage: Therefore, that the world may be benefited by my experience, I have caused my voyages and travels to be printed, which I now present to you, gentle and loving readers, in hopes that the variety of things contained in this book may give you delight. (quoted in Robert Kerr, *A General History and Collection of Voyages and Travels 1824*, quoting Richard Hakluyt's *The Principal Navigations, Voyages, Traffiques and Discoveries of the English Nation*)

This is from what is usually called *Voyages and Travels of Cesar Fredericke in India*, although the complete title is *The Voyage and Travel of M. Cesar Fredericke, Merchant of Venice, into the East India and beyond the Indies: Wherein are contained the Customes and Rites of these Countries, the Merchandise and Commodities, as well of Golde as Silver, as Spices, Drugges, Pearles, and other Jewels.* Translated out of Italian by M. Thomas Hickocke, the book by this real life merchant of Venice runs on to several volumes, as one might expect, just from the length of the title. No point in wasting a long title on a short book.

Writing back in the 1560s, M. Caesar Fredericke observed:

There are many Merchants that stand by at the making of the bargain, and because they shall not understand how the jewels be sold, the Broker and the merchants have their hands under a cloth, and by touching of fingers and nipping the joints they know what is done, what is bidden, and what is asked. So that the standers by know not what is demanded for them, although it be for a thousand or ten thousand ducats. For every joint and every finger hath its signification. For if the merchants that stand by should understand the bargain, it would breed great controversy amongst them.

The oldest known city in Burma is not Ava, Pegu, or Pagan, but Tagaung, which lies north of Mogok on the Irrawaddy River (and probably constructed by migrants from India). It is not old as some cities go, but probably dates from around 200 BCE. It is said that a large ruby mined near there was given to the Kalyani princesses of India. The city was destroyed in the late second century BCE by Chinese invaders and later resettled by Pyu settlers.

These people mysteriously disappeared, but they left behind not only a great many buildings and temples, but a huge stock of precious metals and jewels—jade, amber, and pearls. But no sapphires or rubies. Whether they did not know these stones or did not value them—or whether they valued them so highly that they took them with them, we may never know. The modern city lies on the east bank of the river, but the ancient city was on the other side.

As for the mysterious Caplan or Kyapyin or Capelle, Barbosa, adding yet another name or spelling, wrote:

> And yet further inland beyond this city [Ava] and Kingdom there is another Heathen city with its own King, who nevertheless is subject and under the lordship of Ava; which city or Kingdom they call Capelam. Around it are found many rubies which are brought in for sale to the Ava market, and are much finer than those of that place.

The first Englishman to reach Burma was the London merchant Ralph Fitch in December 1586. Fitch was one of a party of four Englishmen to visit the region, and the only one of them to return alive to England. Although primarily a merchant, surely Fitch's motives for travel were not solely monetary. Some of his excitement at setting off on his voyage shines through in his own words:

> In the yeere of our Lord 1583, I Ralph Fitch of London marchant being desirous to see the countreyes of the East India...did ship my selfe in a ship of London called the Tyger wherein we went for Tripolis in Syria and from thence we tooke the way for Aleppo which we went in seven dayes with the Carovan. ("The Voyage of Master Ralph Fitch, 1598–1600")

He was arrested in Goa, the Portuguese holding in India, where the Inquisition was boiling. Undoubtedly Fitch and his Protestant companions would have been burned alive; however, an English Jesuit priest, Fr. Thomas Stevens, literally saved their skins and arranged bail for them. One of the four travelers promptly decided to become a Jesuit himself "partly out of fear," as he admitted. But not Fitch, who continued on to see Bengal and Burma.

Fitch was tremendously impressed. "Pegu is a citie very great," he wrote.

> In the olde towne are all the marchants strangers, and very many marchants of the countrey. All the goods are sold in the olde towne which is very great, and hath many suburbs round about it, and all the houses are made of Canes which they call Bambos....In the newe towne is the King and all his Nobilitie and Gentrie. The streets are the fairest that I ever saw, as straight as a line from one gate to the other, and so

broad that tenne or twelve men may ride abreast through them....The houses be made of wood, and covered with tiles. The king's house is in the middle of the city, and is walled and ditched round about: and the buildings within are made of wood very sumptuously gilded, and great workemanship is upon the forefront, which is likewise very costly gilded. And the house wherein his Pagode or idole standeth is covered with tiles of silver, and all the walles are gilded with golde. ("An Account of Pegu in 1586–1587")

He also spoke of the gemstone center, Caplan. "Caplan [Kyapyin] is the place where they find the rubies, sapphires, and spinels," he continued. "It standeth six days journey from Ava in the Kingdom of Pegu. There are many great high hills out of which they dig them. None may go to the pits but only those which dig them." As we see, Caplan, according to this account, was six days from the old capital Ava, and Ava, according to di Conti's account, was "two days" from Pegu, while Di Varthema claimed Caplan was "30 days" from Pegu. Perhaps it all depended on the weather. Or perhaps someone's odometer was a little off. The actual measurement used was the *cos*. The only problem with the *cos* was that it was defined as the distance from which one could hear a cow bellow. Since cows bellow at varying volumes and peoples' ears are differently attuned, the *cos* was not particularly satisfactory in this regard.

Fitch was astonished by more than the wealth he found:

In Pegu, and in all the countries of Ava, Langeiannes, Siam, and the Bramas, the men wear bunches or little round balls in their privy members: some of them wear two and some three. They cut the skin and so put them in, one into one side and another into the other side; which they do when they be 25 or 30 years old, and at their pleasure they take one or more of them out as they think good.... The bunches aforesaid be of divers sorts: the least be as big as a little walnut, and very round: the greatest are as big as a little hen's egg: some are of brass and some of silver: but those of silver be for the king and his noble men. They were invented because they should not abuse the male sex for in times past all those countries were so given to that villainy, that they were very scarce of people.

In the earliest days, the Burmese devised an interesting way to conduct the trading of jewels in the open marketplace in such a way that only the buyer and seller were aware of the actual price. They covered their hands with a cloth and used pre-arranged manual signals. Bystanders might rubberneck all they liked, but it would be to no avail.

Caesar Fredericke had nothing but praise for the honesty of the Burmese merchants, too. (The merchants of Venice are well-known not to be so admired.) "It is a thing to be noted," he commented, in the buying of jewels in Pegu that he that hath no knowledge shall have as good jewels, and as good cheap, as he that hath practiced there a long time. There are in Pegu four men of good reputation, which are called Tareghe, or brokers of jewels... through the hands of these four men pass all the rubies: for they have such quantity, that they know not what to do with them, but sell them at most vile and base prices...[W]hen any merchant hath bought any great quantity of rubies, and hath agreed for them, he carryeth them home to his house, let them be of what value they will, he shall have space to look on them and peruse them two or three days: and if he hath no knowledge in them, he shall always have many merchants in that city that have very good knowledge in jewels; with whom he may always confer and take counsel, and may show them unto whom he will; and if he find that he hath not employed his money well, he may return his jewels back to them who he had them of, without any loss at all. (Penzer)

However, as pleasant and easy as the trading habits of the old-time Burmese seemed to be, they had their dark side. If the jewel brokers failed to pay up, according to Fitch, the merchant could "take wife and children and his slaves, and bind them at your door, and set them in the Sun, for that is the law of the country."

The first European to actually visit the mines in the Mogok area and to write about them was the Portuguese (or possibly Italian—he at least published his memoirs posthumously in that language) Jesuit missionary, Fr. Giuseppe d'Amato, who lived in the area from 1784 until he died. His work was translated for publication in the *Journal of the Asiatic Society of Bengal* in 1833. His account is quite short, but provides invaluable information to gem historians.

His observations agree with some of the details of the traditional and mythological accounts of the rather ominous topography of Mogok. He wrote: "It is surrounded by nine mountains. The soil is uneven and full of marshes, which form seventeen small lakes, each having a particular name. It is this soil which is so rich in mineral treasures."

Fr. D'Amato, who included a careful discussion of rubies and trading practices in his day, said primly, "I have avoided repeating any of the fabulous stories told by the Burmans of the origin of the jewels of *Kyatpyin* (Capalan)." This is really too bad, as many people would give a great deal to hear them.

D'Amato also mentioned, rather offhandedly, "There is another locality, a little to the north of this place called Mookop [undoubtedly Mogok] in which also abundant mines of the same precious gems occur." Although the area was pretty swampy, he noted that:

The ground which remains dry is that alone which is mined, or perforated with the wells whence the precious stones are extracted. The mineral district is divided into 50 or 60 parts, which, beside the general name of "mine," have each a different appellation.

The miners, who work at the spot, dig square wells, to the depth of 15 or 20 cubits, and to prevent the wells from falling in, they prop them with perpendicular piles, four or three on each side of the square, according to the dimensions of the shaft, supported by cross pieces between the opposite piles. [This method, known as "twinlone," is still used.]

When the whole is secure, the miner descends, and with his hands extracts the loose soil, digging in a horizontal direction. The gravelly ore is brought to the surface in a ratan basket raised by a cord, as water from a well. From this mass all the precious stones and any other minerals possessing value are picked out, and washed in the brooks descending from the neighboring hills.

Besides the regular duty which the miners pay to the Prince [saopha], in kind, they are obliged to give up to him gratuitously all jewels of more than a certain size or of extraordinary value. Of this sort was the *tornallina* [unclear reference—perhaps touramaline or even zircon] presented by the Burman monarch to Colonel Symes. It was originally purchased clandestinely by the Chinese on the spot; the Burmese court, being apprized of the circumstance, instituted a strict search for the jewel, and the sellers, to hush up the affair, were obliged to buy it back at double price, and present it to the king...The precious stones found in the mines of *Kyat-pyen*, generally speaking, are rubies, sapphires, topazes, and other crystals of the same family (the *precious corundum*). Emeralds are very rare, and of an inferior sort and value. They sometimes find, I am told, a species of diamond, but of bad quality. The digging in a square area is still practiced, although there is a new "wash plant" in Mogok that uses high pressure hoses to clean off the accumulated dirt of the stones.

The British were well aware of the presence of rubies in Burmese holy places. Indeed, it is the main reason why more than 1,000 pagodas were desecrated or destroyed by British troops at the end of the Third Anglo-Burmese War. It is an old saw that a king would be ruling at Mandalay today if it had not been for the irresistible lure of rubies.

In the sixteenth century, the Burmese monarchy took control of the mines and commanded that all rubies over a certain size (and valued at 2,000 rupees) were automatically the property of the Burmese kings, some of whom seemed to think their elevated position freed them from the ethical precepts of their religion. Failure to pass along the good rubies was a crime punishable by torture and death.

Not happy with the low production, in 1780 the Burmese king Bodawpaya sent thousands of captives from the Manipur war to work in the mines in Mogok until they died. This didn't take very long, and mines became a de facto penal colony. Two years later, in 1783, Bodawpaya extended the tract boundaries to encompass Mogok, Kyatpyin, and Kathé.

Even Europeans were pressed into the mining business. In 1830 a runaway English sailor employed by the Burmese King Phagyidoa was sent off to blast rock at a royal ruby mine at Tapambin. Perhaps he died in the attempt. Or perhaps he decided that he liked mining no better than he did seafaring, for he was never heard from again.

THE RUBY WARS, OTHERWISE KNOWN AS
THE ANGLO-BURMESE WARS

Where there are riches there will be wars. The Anglo-Burmese Wars began in 1824, when the British first landed in Rangoon. There were three of them altogether. The first one (1824–1826) was won by Britain.

The Burmese first came into contact with the British when the Burmese conquered the small principality of Assam and turned it into a client state. The British retaliated and in May 1824 showed up in Rangoon ready to conquer. However, that was a few days before the monsoon hit and before long entire country turned into a fetid quagmire. Fever broke out among the troops, killing hundreds. However, luck was on the side of the British when the Burmese general was killed by a stray cannonball. The Burmese then gave up and let the British sail triumphantly up the Irrawaddy River to Ava, the ancient capital, where the king sued for peace and let the British have Assam. However, there was something odd about the treaty. The "Treaty of Yandabo" ceded Arakan, Assam, and Tenasserim to the East India Company, a commercial company, not to England, a sovereign nation. However, the East India Company was more than a tea and silk trading enterprise. For a considerable time, it was the de facto government of a lot of places, including India, Burma, Malaya, Singapore, and Hong Kong, until it was dissolved in 1858.

The Second Anglo-Burmese War began when the Burmese King Pagan Min (ruled 1846–1853) tried to gain control of the Arakan territory. The British beat the Burmese again, and Pagan Min was deposed, leaving his brother Mindon Min to take the throne in Ava in 1853. Things went well for quite a while. Mindon Min was a fairly benevolent dictator who tried to accommodate the British, something that was never easy.

Mindon Min not only gave generously to the famous old Shwe Dagon Pagoda, as mentioned earlier, but also built a new palace, in response to a Buddhist prophecy. He finished it in 1859 and named it the Heavenly Abode. This sumptuous palace contained a Great Hall of Audience or Myey Nan Pyat-that, with ten rows of gold-gilded teak columns and a series of seven vermilion roofs (which represented the seven-tiered earth). The only woman officially allowed in was the queen, although other court ladies could look on from behind a screen. There was also a Chamber of the Lion Throne (supposedly the center of the universe), and a room whose walls were covered in glass mosaics. The Lion Throne was made of "yamanay" wood and gilded with 24-karat gold leaf. Just above, there were the Nine Noble Gems: the ruby for glory, the diamond for honor, the pearl for grace, red coral for greatness, the zircon for strength, the sapphire for adoration, the cat's eye for power, the topaz for health, and the emerald for peace. On the lintel were Bodhisattvas, or heavenly Buddhas, eight on each side, standing on lotus blossoms, the sacred flower of Buddhism. A rabbit and a peacock, who represent the moon and sun respectively, make an appearance as well. There is also a carving of a lion fighting with an elephant.

King Thibaw (reigned 1875–1885), the last king of Burma's Third or Alaungpaya Dynasty and really the last king of Burma, had other thrones, too, besides the lion one, each made of lustrous wood. Officially there were nine thrones, but since there were two (identical) Lion Thrones, it is usual to speak of the Eight Thrones. The number eight is a sacred one in Buddhism, and the Burmese king was said to have Eight Special Virtues or Powers.

The other thrones, each made of lustrous wood, included the Hintha Bird Throne (the hintha bird symbolizes serene authority), the Peacock Throne (representing the love of the Buddha and not to be confused with the Indian/Persian Peacock throne), the Lotus Throne (for the Queen), the Bee Throne (wisdom), the Elephant Throne (longevity and sovereignty), the Conch Throne (wealth), and the Deer Throne, which symbolized prosperity for the nation. Technically, there were nine thrones, but since there were two identical Lion Thrones, it is usual to speak of the Eight Thrones. The number eight is a sacred one in Buddhism, and the Burmese king was said to have Eight Special Virtues or Powers. Each throne weighed a ton, and was more than 34 feet high, with the seat itself being six feet high, and all were carved simultaneously. Legend says the carvers were given tools of gold and silver with which to work. This doesn't seem practical, so it can be supposed either that only the handles were gold and silver or else that the tools were simply gilded. Or that the chroniclers lied.

Most of the king's thrones were lost when Mandalay was bombed during World War II, although the Royal Lion Throne is on display in National Museum in Rangoon, as well as miniatures of the others.

Mindon Min did his best to try to get along with the British (and the French) who were vying for control of the area. In 1872 he sent one of his best diplomats to see if he could negotiate sovereign status for Burma, even though the British had boots on the ground all over the southern part of the country. The envoy made it all the way to England, even to the audience chamber of Queen Victoria herself. The Queen got a lot of nice presents from the visitors—including a ruby bracelet and belt. The Crown Prince got a ruby-studded gold belt. However, as one might expect, no official promises were made.

The delegation next tried their luck with the French, where they met with the foreign minister and French president (who probably also received some jewelry). The French and Burmese concluded two treaties, both of which benefited the French and neither of which provided the Burmese with arms, which is of course what they were really after.

Before the treaties could be ratified, rebellion broke out in Burma. Mindon Min escaped, but his heir designate was assassinated. King Mindon escaped death in a remarkable way. It happened that when the man assigned to kill him actually ran into him face to face, he couldn't go through with it. In fact, he dropped to his knees, confessed all, and offered the king a piggy-back ride escape to the barracks of the royal guards. The king himself became very ill a few years afterwards and died in 1878 at the age of 70. This was the death-knell, not only for Mindon Min, but for Burma.

Mindon Min's death was a sad affair in more ways than one. Before his death there had been a power struggle between two of his queens, his Chief Queen, and a lesser wife who tried to seize domestic control of the dynasty. It didn't work, but Mindon was nervous enough about the whole affair to bestow upon her a new name—"Lord of the White Elephant." Although you might think she would have been satisfied with her new title, she wasn't. After the king died she managed to get her two daughters (Supayagyi and Supayalat) married to Thibaw (ruled 1878–1885), a rather doltish fellow, but son of another royal wife. He was the last king of the third or Alangpaya Dynasty Supayagyi turned out not to be of much help in her mother's plans and soon disappointed her completely by dying, but the other daughter was truly her mother's own. She was vicious, petty, and cruel, and she soon became Thibaw's chief wife.

Thibaw himself might have been content to stay at home and play with his rubies. It would have been better for everyone if he had. As a result of his various royal thefts, he was supposed to have the finest collection of rubies in the world, including a 100-carat rough stone found on Pingtoung Hill (Pingu Taung) near Mogok. It was presented to Thibaw by Oo-dwa-gee, the governor of the ruby mining district.

In 1885, Thibaw appointed a governor, whose job it was to collect 16,000 pounds of rubies per year to give to the kings. This did nothing to endear the king either to his own people or to the English, who wanted the stones themselves. Not for nothing did Rudyard Kipling (1865–1936) write:

And they were stronger hands than mine
That digged the ruby from the earth
More cunning brains that made it worth
The large desire of a king.

For Thibaw to acquire the Burmese throne meant killing off his rivals. The killing that had to be done was not a pleasant affair, nothing easy and clean like a firing squad. Burmese tradition demanded that royal blood could not be shed. The simple solution was to stuff all Thibaw's hapless half-brothers and sisters into gunny sacks and then have them beaten to death. Some sources say the victims were stomped by elephants. (There were a lot of brothers—about 70 of them. Thibaw's dad, King Mindon, was a busy man.) The sacks were made of red velvet, so you couldn't see the blood. Actually, Thibaw's wife Supayagyi and his mother-in-law Lord of the White Elephant did most of dirty work. (She may have been rather closely related to her husband. She may have been his half-sister.)

All this occurred in February 1879 in Mandalay. The story goes that at the point of execution, one of the royal princes turned to his brother, who was begging for his life, and said sternly, "Brother, it is not becoming to beg for life. We must die, for it is the custom. Had you been king, you would have given the same order. Let us die, since it is fated we must die." The Burmese minister also defended the killing to a shocked world, announcing that since the king was an independent sovereign, he could do just as he pleased, a substantial gap in logic.

This brutality was enough to make even the British nervous. They were not particularly interested in the human rights of the Burmese royal family, but they did care about the rubies, especially when they heard rumors that Thibaw had ceded the rights to the ruby mines to the hated French.

Thibaw determined to outwit the British, a plan doomed to fail. It was difficult for Thibaw to outwit anyone. In 1883 he sent another Burmese mission to Europe in an attempt to get the earlier treaties ratified. The delegation met with the French Prime Minister Jules Ferry. The British made a huge fuss about it, claiming that there were dangerous secret clauses in the treaty that would be harmful to British interests (there weren't). Negotiations were stalled and Ferry fell from grace, ending the talks altogether.

The nervous British moved their residents out of the immediate vicinity, and then decided to take over the country. The war actually began in 1885 when the liberal Gladstone government fell and a group of conservatives took over (the precipitating cause was a proposed tax on beer and liquor). The new Secretary at the India Office was the imperialist Randolph Churchill, who hypocritically wrote, "The general interests of humanity are infringed by the continued excesses of a barbarous and despotic ruler." Kipling celebrated the affair in cheerful, if jingoistic verse:

> On the road to Mandalay,
> Where the old Flotilla lay,
> Can't you hear their paddles chunkin'
> From Rangoon to Mandalay?

As the British arrived, the king thought it might be an excellent idea to distract the mind of the court with dancing, so he gave a ball (although he was smart enough to pack up the royal elephants with treasure and send them off).

However, it is not true, as Brit-bashers insist, that the British had no higher motives and were interested in only Burmese jewels. On the contrary, they also had their eye on Burmese teak, gold, silver, tin, and tungsten. In fact, the entire British invasion was rationalized by a fuss about teak logs and the taxes thereon.

The British attacked with three brigades of Indian soldiers and a regiment from Liverpool. Thibaw's army quickly lost the "battle" at Minghla south of Mandalay, and the British surrounded the palace to prevent his escaping and thus prolonging a hopeless conflict. On November 27, 1885, the British arrived at Mandalay. While General Prendergast managed the troops surrounding the palace, British Colonel Sladen politely followed the established rules of war and entered the palace via the southern gate—the one reserved for foreigners. Thibaw sat nervously on this throne in the Grand Hall of Audience, probably wondering if he was going to be thrown into a gunny sack and beaten to death. His wife, who was there also, was probably wondering the same thing. The King's minister, the Kinwun Mangyi, escorted the conquering British colonel into the Great Hall, where the King begged to be allowed not only to live but also to keep his prize ruby, the Nga Mauk. This is the stone discovered back in the reign of King Pindale (1648–1661). Sladen ostensibly agreed, but asked just to see the precious stone, which was secreted in a jewel-studded betel box. (Some sources claim he had seen the gem earlier, and thus knew where it was.) He then suggested to the stupid Thibaw that the Nga Mauk would be much safer with him, and he stuck it in his pocket. No one has seen it since.

When the king later asked for the jewels back, he was told that the colonel had gone back to England. Sladen died in 1910, and no one knows what happened to the ruby. Or if they do, they are not telling. (It is perfectly possible that this great stone was a spinel, anyway. There's no way to test it now.)

The British were not very clever about grabbing their share of the loot. Apparently, they were lounging around the palace of the beaten Burmese king Thibaw, watching a plethora of Burmese ladies darting around the palace with urns and bowing obsequiously.

Even though they had "sealed" the palace, the British acceded to the request of Thibaw's minister to allow Queen Supayalat's ladies-in-waiting to come and go as they liked, ostensibly to serve tea and so on. Presumably this was a matter of honor and courtesy. However, there was more in those urns than tea. In fact, most of the Burmese royal rubies disappeared during the annexation, along with almost everything else in the palace, including the sacred dynastic images. The British, by the way, claim that it was the ladies-in-waiting, not Colonel Sladen, who walked off with the Nga Mauk. It probably doesn't matter, anyway, since whoever took it kept it. I should say that there is an allegation by Thibaw's grandson that the so-called Chrismore Ruby in the United Kingdom is in fact the Nga Mauk.

So, while the Brits were searching the closets and chests for signs of rubies, they ignored the women. Besides, it wouldn't be cricket to search them. It took the women only a day to clean out the entire palace, tucking vast numbers of the priceless gems into their sarongs. Legend says that they got away with the Nga Boh ("Dragon Lord") Ruby, an extraordinary stone found in Bawbadan, weighing 44 carats in the rough, and later cut to 20 carats, as well as another rough stone recut into a 70 carat gem; this stone was first given by the finder to King Tharawadi (1837–1846), probably under pressure.

Still, the British did manage to scrape up enough leftovers to be worth about a million pounds, including Thibaw's royal conical crown of beaten gold (with eight stones missing). King Thibaw supposedly had 20 other crowns, which no one claims to have seen, although the Burmese maintain that Colonel Sladen made off with them. It is certainly possible. The British got a bunch of emeralds too, which, of course, were not mined in Burma, but in Colombia. They picked up the King and Queen's robes, the royal sword, betel boxes, as well as ruby encrusted slippers, possibly the prototype for Dorothy's in the *Wizard of Oz*. And some dinner plates. Perhaps some of the missing loot was the king's ceremonial outfit that consisted not only of his crown, but also a pair of massive golden wings flying out from his shoulders. He was supposedly so top-heavy that he needed to clutch at a balustrade to keep his balance. A good deal of this stuff ended up in the Victoria and Albert Museum. There were lots of rumors swirling about that many of the precious stones were buried outside the palace, giving rise to a frenzy of digging, although no stones were ever found.

The king was sent into exile in India in a cart drawn by bullocks, along with his family, and as it turns out, a great quantity of precious gems. Some say the king packed them off in his long hair. (Like his luckless brother, he ended up begging for his life, although his queen was defiant, with her child clinging desperately to her side. The British were more merciful than Thibaw himself had been.)

It is known that Queen Supayalat escaped with a handkerchief full of her personal jewels, dropped it getting on to a steamer, and had it returned by an honest soldier. Most of the recovered jewels from the Royal Palace at Mandalay are currently at the Indian Museum in London. However, it's obvious from looking at them that someone made off with the best of the lot. Most of the ones on display are small and of imperfect quality.

As for the palace itself, the British converted it into an officer's club, and renamed it Fort Dufferin. They desecrated the place by walking around inside with their shoes on. This has been a point of contention from the earliest days of the British presence in Burma. The Burmese had wanted the British to take off their shoes, and the British refused, and so on. It became quite a diplomatic scuffle. It even got a name: "The Great Shoe Question." (Shoes did not again play an important part of international hoopla until the Imelda Marcos regime during the following century.) The palace itself was totally destroyed in 1945, during World War II. Only one wall remains.

AFTER THE WARS

On January 1, 1886, Britain formally annexed Upper Burma. For a few months, for the first time in centuries, there was a melee of free mining in Mogok, and stones were sold without restrictions by the people who actually found them.

However, this pleasant state of affairs was not to last. On December 26 of that same year, British troops reached the Mogok area, and on January 27, 1887, entered the town of Mogok itself.

The British had no intention of taking over the country in order to let the peasantry run it. Edwin W. Streeter, a London jeweler, along with Col. Charles Bill and Reginald Beech (for a bid of 400,000 rupees), obtained the concession for the mines from the India Office. (Another company, Gillanders Arbuthot of Calcutta, had also been interested.)

George Skelton Streeter (E. W. Streeter's son), Charles Bill, Reginald Beech, and engineer Robert Gordon came along. It was not until 1887, however, that a systematic description of the deposits was made, when on

January 10, 1888, C. Barrington Brown, a trained geologist, was sent to Mogok by the Secretary of State for India to check them out and write up a proper geological description of the area. He noted that there were extensive deposits of ruby-bearing alluvial matter as well as spinel, garnet, tourmaline, and rock crystal. A thorough geological survey was undertaken between 1887 and 1893 by Colonel J. R. Hobday.

The squabble over the rights to the mines eventually involved members of Parliament itself, and the public caught ruby fever. In 1889, the Streeter syndicate joined with the Rothschilds of France to form Burma Ruby Mines Ltd., which was floated on February 26. Ordinary shares rose to a 400 percent premium, and the whole issue was sold out in minutes. The British government was to receive 300,000 pounds per year plus 30 percent of the take from the poor native miners who were doing the actual work. This, of course, led to an immense amount of smuggling, and it was simply impossible to regulate the prices of rubies. Even today many Mogok rubies are sold directly outside the ruby mines at 5:00 A.M., a trade largely unregulated even by the Burmese military, which is supposed to be looking out for such things.

The proprietors of the mine commissioned some geological surveys that revealed that the richest source for rubies lay directly under the village of Mogok. With the serene aplomb born of those who thought it their right to rule, the British simply brought in some heavy equipment, dismantled the town, relocated the entire city to a nearby site, and started digging. Their plan was to strip everything, they said, "from earth to bedrock." (They said they did it so that the town wouldn't wash away during the rainy season.) They also built a 400 kilowatt hydroelectric station, five washing mills (processing thousands of tons of rock every day), and a mile-long tunnel designed to handle seasonal runoff dug into the solid rock. Then the soil was trucked off to a washing mill and inspected.

It was during this period that most of the museum-quality rubies of the world were acquired. (The Maharajahs had been simply content to set them in their turbans, without bothering to sell them.)

After the development of the synthetic ruby in 1908, the natural ruby market became extremely volatile. There was widespread panic and depression (both economic and otherwise), and the price of natural rubies swung wildly. Then there was the First World War. The mining company surrendered its lease and everything went back to the old way of doing things, at least until after World War II, when modern equipment was introduced.

Therefore, despite an auspicious start, the profitability of Burma Ruby Mines Ltd. was volatile, and the company went bankrupt, partly due to its overreliance on heavy machinery such as electric generators, instead of the low-tech (and low-cost) mining methods of the natives. It didn't help that the British insisted upon importing their entire way of life—and it wasn't cheap. The company also insisted upon digging at non-profitable mines in Kyapyin area, famous of yore, but worked out for centuries.

The political climate also became increasingly tense. In 1920 the Young Men's Buddhist Association (which did indeed begin as the Buddhist imitator of the YMCA) began campaigning for dominion status and home rule. Political parties such as DoBana Aseyone did likewise.

There was a rebellion in 1930, led a by teacher named Saya San. He claimed that he was immune to British bullets because he had special tattoos all over his body, and he also carried ruby talismans to ward off all danger. It worked for a while, but the British finally caught up with him and killed him anyway. Eventually, however, an agreement for self-government was reached, and in 1937 Burma had its first modern Prime Minster. Equally important form the Burmese point of view was that Burma was separate from India (the British seemed to have trouble telling the countries apart and lumped them together). But the currency remained the rupee.

Another factor leading to the decline of the ruby market was the Depression. Although it took a while for the worldwide depression to hit Burma (which is rather far from the economic powerhouses of the planet), 1929 turned out to be a particularly awful year for rain. There may have been drought in Oklahoma, but there was no lack of rain in Burma. The relentless downpours flooded the tunnel the British had dug and destroyed most of the electrical power equipment. However, Ruby Mines Ltd. struggled on for a couple more years before finally giving up and returning its lease to the Burmese government in 1931. (There are some rumors that one of the real culprits in the whole business was the De Beers diamond cartel, who some believe engineered the failure because they saw rubies as a possible competitor to diamonds. It is hard to see how the De Beers Company was responsible for the rain or stock market crash, or even for the synthetic ruby, but anything is possible.) Burma Ruby Mines Ltd. was liquidated in June 1934 and replaced by a new company primarily owned by Arthur Henry Morgan, John Francis Halford-Watkins, and Charlies Lewis Nichols, and run by a board of directors. Despite the finding of some excellent stones, the Depression made it hard to sell any of them.

Despite its eventual demise, the Burma Ruby Mines company did have some excellent gem discoveries on its track record. One of the first major finds of the occurred in 1899, in the form of a 77 carat rough which was valued at 26,666 pounds. The whereabouts of this ruby, if indeed it was one, are currently not known.

The last big find was probably the so-called Peace Ruby, a 42 carat rough discovered in Mogok on June 30, 1919. The World War I peace treaty was also signed this day, hence its name. It was in the form of an "irregular hexagonal prism" with a flat apex and of spectacular color. The stone was sold to an Indian gem dealer named Chhotalal Nanalal for 27,000 pounds (about 654 pounds per carat—a record at the time), and was subsequently recut in Bombay to 25 flawless carats. It was afterwards sold in America, but like so many other notable rubies, has since disappeared from view.

WORLD WAR II AND AFTER

After the British returned the lease, mining methods went back to the old ways under private ownership—at least briefly. Then came World War II and the Japanese invasion of Burma. The Japanese bombed Rangoon on December 23, 1941, and the British families living there, including the managing director of the mines and his staff, had to flee to India. During this period and its aftermath (1942–1945), almost no mining was done.

At one point the British naively believed that the Burmese would join them in their fight against the Japanese. However, they soon realized their mistake. While the Burmese were not crazy about the Japanese, they hated the English even more, and probably thought that the Japanese wouldn't be any worse. They weren't. But they were just as bad.

When the Japanese conquered the Mogok region, they took everything with them they could carry, including most of the mining machinery. The mines returned to their previous, unimproved condition. The Japanese also ruined the rest of the Burmese economy, declared Burmese currency illegal and created rampant inflation. Foreign capital disappeared and the only people with access to the professional health care system were members of the Japanese army. The natives had to rely on traditional folk methods of healing, some of which involved rubies. Luckily for the Allies, Burma, and the civilized world, the Burmese Independent Army revolted against the Japanese and helped the 14th British army liberate Rangoon in May 1945. The British were back in power, but their control over the country was considerably more enlightened. After the war, native mining resumed. But not for long.

Ruby Mines, Ltd., resumed operations, despite the dreadful conditions of the mine. The only properties left were the powerhouse and the generators (not in working condition), one main office, and some high tension lines. In addition, prospects for gem sales were still bleak, as much of Europe and Asia had been ruined by the war.

To makes things worse, almost the entire government of Burma was assassinated on July 19, 1947. Only one senior minister, U Nu, survived, and he formed a new government in September of that year. The exact time for independence was based on an astrological determination, which deemed the proper time to be January 4, 1948—at 4 o'clock in the morning. As one might expect, the good times didn't last long. Communists and various ethnic groups vied for control of the country. (Despite the turmoil, the ruby mines had an excellent decade between 1948 and 1958.)

In March 1962, a military coup under General Ne Win took place. (The name Ne Win means "brilliant like the sun." He named himself.) Parliament was dissolved, the constitution put on a shelf, and a socialist agenda put into place. Private businesses were nationalized, and the country went into an isolated, vicious dictatorship, similar to that of the nineteenth-century Burmese kings.

Ne Win also moved his troops into the ruby-rich Shan States. In 1969, the Burmese Ministry of Mines banned further exploration and mining of gems, and essentially nationalized Burma's gem mines, with the former owners receiving no compensation. It didn't seem to matter, for little of importance was produced. The program was called the "Burmese Road to Socialism," a stated objective that does not, to say the least, encourage the entrepreneurial spirit, and is nothing at all like "The Road to Mandalay." The slogan was rather promptly changed to the "Burmese Road to Poverty" by members of the public brave enough to speak up. Burma had once been one of the most prosperous countries in Southeastern Asia. After the military takeover, it became one of the 10 poorest in the world. Ne Win officially resigned from office in 1988 following a series of demonstrations, but his replacement, Sein Lwin, affectionately known as "the Butcher," can hardly be said to be an improvement.

Previously issued ruby mining rights were revoked by the army, almost completely isolating the area economically. No foreign journalists were allowed to visit. This had the effect only of increasing the smuggling—soon more Burmese rubies were for sale in Bangkok than in Rangoon. (In 1988 thousands of Burmese citizens were gunned down by the authorities for daring to voice their opposition to the government. However, the deaths were not in vain, as they resulted in some liberalizing measures.) In 1990, the ban was partially lifted. This was the year that ordinary people were once again allowed to mine stones, so long as they handed over the best gems to the government, in time-honored fashion. Anything over a certain value was to be sent to Rangoon for auction, which could be avoided only by paying a 20 percent tax on the stone.

Also in time honored fashion, the finders decided to smuggle the stones out of Burma into Thailand to get a better price. Four men involved in the scheme were caught and sentenced to life in prison. One man, Ko Win Bo, a former Army captain, was released for the first time in 1997, and later became a political prisoner while he was working for the release all political prisoners. He was flogged to death in 2000 by 23 prison employees for defending prisoners' rights.

In 1988, widespread demonstrations resulted in the ousting of the general, who was replaced by Daw Aung San Suu Kyi, the daughter of Aung San, a charismatic and beloved Burmese leader, who had helped lead Burma to independence in 1947, but who was then assassinated. Suu Kyi (pronounced Soo Chee) herself was soon placed under house arrest by the military dictatorship, was released after international pressure, and finally re-jailed. Suu Kyi won the Nobel Peace Prize in 1991. Since she was not allowed to leave the country, her oldest son picked it up for her. Suu Kyi has remained on and off house arrest for many years, but when offered a chance to leave the country to attend her dying husband in England, she refused, fearing she would not be allowed back in. For most people that might be good thing.

The Burmese mining and ruby trading policy has been unstable, to say the least, changing almost from month to month. For a while, Burmese citizens were allowed to apply for two-year mining licenses, as long as, in accordance with ancient custom, they gave all the best stones to the Burmese government. The more things change...

During Burma's "socialist period" even owning a gemstone was illegal, much less trading one. All gem exports were supposed to go through the Myanmar (as the country suddenly became known) Export Import Corporation. Today, the governmental overseeing body for rubies and other precious stones is the Myanmar Gems Enterprise, which sounds more enterprising than it actually is.

It was during the period of the ruby trade ban that the "official" name of the country was changed to Myanmar, and Rangoon became "Yangon." The reason for this is not political but superstitious. Ne Win is famously superstitious. A common Asian belief is that you can stop a run of bad luck by changing a name. (The mischief-making spirits won't be able to find Burma if it becomes Myanmar.) In any case, for those in the gem trade (and also for the United States State Department), Burma is Burma and not Myanmar, and there is no use pretending that it is, no matter what the

current maps may say. The concept of Myanmar rubies simply doesn't reso-
nate. Actually, most of the gem-bearing countries in the region have changed
their names. Sri Lanka was once Ceylon, and Thailand was Siam. And
Burma, even before it became Myanmar, was once Bamar. Luckily, the pro-
nunciation of "Bamar" is practically identical to "Burma," at least for the
British. Try saying them both aloud a few times and you'll see what I mean.

In 1988–1989 a new government was formed, ending the socialist experi-
ment and retracting all previous gem laws. Reforms were instituted, and the
Ministry of Mines was reorganized. The Myanmar Gems Corporation
became the Myanmar Gems Enterprise (sadly the country wasn't allowed to
go back to its real name). And later, in 1990, private enterprise was allowed
back in, and in 1995 joint ventures were welcomed. Individuals were now
allowed also to carry gemstones freely, as long as license was obtained.

THE NEW RUBIES: MONG HSU

In about 1992, large, vividly red rubies starting showing up from a new
Burmese mining area surrounding a village called Mong Hsu, which lies in
the middle of the Shan State, about 150 miles southeast of Mogok, and
170 miles northeast of Mandalay.

The name Mong Hsu means "meeting with satisfaction," and there is an
old story about how it got its name. Once upon a time there was a bachelor
who developed an itch for travel, and his wanderlust led him westward. At
the same time, a spinster living on the other side of the land came to the same
conclusion, and she traveled east. The place where they met and exchanged
stories became Mong Hsu. It is said that when they died, their bodies grew
into the two hills that dominate the place where rubies are now found. Apparently the local inhabitants knew of the existence of the "red sands," as they
called them, but had no idea that they were actually worth anything. By most
accounts the great discovery did them little good, and they went from sup-
porting themselves through agriculture to losing their lands to the mining
interests. If they looked for gems on their own former lands, they were
arrested as trespassers. The farmers ended up having to buy the rice they
formerly could sell. When they ran out of money, they became beggars.

(A few rubies had actually been mined there in the nineteenth century, but
because they were mostly opaque and not easily faceted, the mines were
abandoned. The deposits are of upper Paleozoic marble, probably metamorphosed at somewhat lower temperatures than those at Mogok.)

Mong Hsu stones are opaque in the rough. They tend to be smallish, and
the cores have a rather blue or purplish look, rather than being uniformly

red. They also possess dense rutile silk clouds, and before they are heat-treated, they can easily be mistaken for garnets. Most of those sold in the first open market are quite heavily color-zoned and fractured. The color problem can be removed by heat treatment, which turns them brilliantly red. In fact, it can be safely said that 100 percent of Mong Hsu rubies are heat-treated. Star rubies have not been reported from the Mong Hsu site.

The fracture problem is more recalcitrant. Such stones are usually "healed" by heating them with borax and other chemicals. This melts the whole surface, including fractures. The corundum then redeposits on the surface of the fracture and seals it shut. So you can technically say that the fractures are not filled, but healed with "synthetic ruby." This treatment improves the stone's durability. It is impossible, for the most part, to tell where the "natural" ruby and "synthetic" ruby join. This treatment goes deeper than simple heat treatment, and the buyer is usually not told that what is being sold is not in its original state. Again, I am not talking about color here. Everyone assumes that all rubies are color treated unless guaranteed otherwise. This is altering the stone. To further the problem, most gemologists in Thailand, where the stones are traded, accept this treatment without comment; however, gemologists and traders in the rest of the world object to it. Today heat-treated Mong Hsu rubies are offered in commercial qualities and in sizes between 0.5 and 3 carats.

Mong Hsu rubies can be beautiful, but they can't compare with those from Mogok. While they seldom display the solid inclusions found in Mogok stones, fluid inclusions are common. The fluorescence of Mong Hsu rubies is also different from those of the Mogok mines. Before heating, the cores' fluorescence ranges from inert to light orange to light red, although the outer rims fluoresce an orange-red. After heat treatment, the cores fluoresce orange-red to red, while the rims stay an intense red.

When the Mong Hsu mine was first discovered, enough rubies were unearthed to lower prices worldwide. By March 2002 more than five million carats of Mong Hsu rough rubies were sold, while Mogok stones sat on the shelf (so to speak). However, the Mong Hsu mine is less productive today, although most stones less than three carats are Mong Hsu rubies, and nearly everything traded in Chanthaburi (Thailand) are from Mong Hsu.

This sudden bounty from Mong Hsu pushed the prices down from as much as $20,000 for a single carat stone to a mere $3,000. The Thai border town of Mae Sai became the main smuggling point for these gems. In 1994 the Burmese government reduced the export tax on gemstones to 15 percent in an attempt to cut down on the smuggling, but it didn't work. In 1995, the government closed all ruby markets at Taunggyi, and moved the legal trading down to Rangoon. In 1995, 224 new mining licenses were issued, which jump-started the exploration for new gemstones. Currently there are seven government-operated mining concerns. There are also many joint venture leases with local people. This hasn't stopped the smuggling, of course. About half the rubies of Burma are smuggled across the porous Thailand border for sale there. The government is so tight that only about 60 people are allowed to deal directly with the Burmese government at its annual gem sale. However, this is a policy that changes erratically, so it's always best to check for updates.

However, the totalitarian government of Burma is not a trading paradise; its inconsistent policies make it impossible for merchants to know from one day to the next what will be permitted and what will not. As a consequence, the center of the ruby trade is now Thailand, which has a wonderful cutting and polishing industry that is centuries old.

It is also important to understand that Burma still uses forced labor, the systematic rape of women, and ethnic cleansing as part of its daily business. The ruby is indeed a blood stone for which gem dealers and even governments are quite willing to look the other way. In fact, today Burma is producing more stones than ever.

OTHER LANDS, OTHER RUBIES

While the world's best rubies are famously from Burma, specifically the Mogok region, they are not the world's only source for the precious red stone. Nor is every Burmese ruby a gem worth having. There are bargains and treasures in other lands. In fact, because Burmese rubies have a reputation of being "better," than those from other places, it's comparatively easy to pick up a bargain a stone from a different country of origin.

Thailand and Cambodia

While Burma stands as the world's pre-eminent ruby producer, Thailand has also been a major producer, especially during the time the Burmese market was essentially closed (from the early 1960s to 1991, when the Mong Hsu stones hit market).

Thailand and Cambodia are the only ruby-producing areas whose gems often do not have the needle-like rutile silk inclusions that routinely mark most of the world's rubies. In some cases rutile inclusions are present but are finer than those in Burmese rubies. This reduces the value of the stones. The Thai-Cambodian border is also the only area in which spinels are not found in the same mines as rubies. During times when Burmese rubies were not commercially mined, Thai rubies comprised about 70 percent of the world's production. However, since the re-opening of the Burma market, Thai stones have generally faded from the marketplace, and its cutting factories are dependent on stones from Africa, Vietnam, and Burma. Since 1919 mining has been restricted to Thai nationals.

About 400 miles south of Mogok is Thailand's famed Hill of Precious Stones, where most of its gems were found. The Hill is near the Cambodian border, in the area of Chanthaburi (located near the Gulf of Siam). Rubies are found in lateritic soil, a red residual soil formed by the leaching of silica and enriched with aluminum and iron oxides. This kind of soil is especially common in humid climates and usually tops Plio-Pleisticene rubies

One of the great advantages of Thai rubies is the comparative ease with which they are dug up. Most are found only a few meters deep in basalt rubble. Other important Thai sources were Trat and Borai.

Even though Thai stones are very clear and often of good size, the lack of silky sheen means less light scattering and thus less brilliance. As a result, only those facets where the light is completely *internally* reflected have the desirable rich "ruby red." However, in rubies, color is everything. A few people actually prefer Thai stones, as they are attracted to the "purer" red of these stones. Since Thai stones contain so little rutile, the finest examples have a very bright, clean appearance. This is not to everyone's taste, as most ruby devotees are more charmed by the soft luster of Burmese gems. As with everything else in aesthetics, this is a matter of personal preference.

Another problem with Thai rubies is that most of them contain too much iron (the Chantaburi-Trat mining district in Thailand is loaded with iron). When Thai rubies where taking shape about 150 million years ago, they incorporated some iron atoms into their structure, which has the unfortunate result of lending many of them a brownish cast. The iron also represses fluorescence; Thai rubies are just barely fluorescent, even under ultraviolet light. However, the brown effect is something of an artifact. In actual fact, Thai stones are the most deeply red of all rubies, some of them even bordering on black so that they rather strongly resemble garnets, which isn't good either. The Thai word for ruby is *taubptim,* which also means pomegranate, another unhelpful comparison. The red of Thai rubies is nothing like the red of a pomegranate.

It is sometimes believed that rubies from the Thai-Cambodian border have more "purple" in them than Mogok stones. This isn't really true. In fact, the opposite is the case; Thai rubies can be of an exceptionally pure red. In fact, stones from the Thai-Cambodian border may possess a "purer" color than the Burmese variety, but lacking the fluorescence of Burmese stones, they also lack its emotional power. The objection to Cambodian or Thai rubies may also have more to do with saturation and darkness than any perceived "purple." Technically, purple is a hue position, an intermediate hue between red and violet. In gems, "purple" usually refers to a low-saturation red (grayish) characteristic of Thai stones compared with the higher saturation (but in actuality more purplish) red of Mogok stones. Much of this overtone can be eliminated by the heating process, however.

Mineralogically speaking, one of the most characteristic and intriguing inclusions of Thai rubies is the appearance of one or more of the so-called Saturn inclusions, which consist of a core with rings around it. Another is a "fingerprint" pattern that either cuts through the center of the core or lies against it. And even Thai rubies of high clarity often appear in flat crystals that result perforce in a shallowly cut stone.

Nowadays, Thailand is probably more famous for cutting rubies than for finding them. In fact, nearly every Old World ruby originally from Cambodia, Kenya, or Tanzania, and most of those from Burma, eventually ends up in Bangkok to be heated, cut, and polished. This is not always a sophisticated affair. For the most part, the Thai market is very simple—with most stones just sold on tables by Chinese intermediaries.

Much of what is true of Thai stones is also true of most Cambodian rubies (the rubies from this region straddle the border between the two countries). That inveterate explorer Tavernier reported that Cambodia also produced gold, sapphires, and topazes. Cambodia is rich in other minerals as well. Some of the rubies he referred to as "spinelles," by which he did not mean the mineral spinel (which is not found in the region), but a light-colored ruby. Again, the distinction between the various red gems was quite uncertain at the time. Tavernier also professed a great interest in another Cambodian commodity—rhubarb. Apparently, the Cambodian variety of this herb didn't go bad as quickly as other Asian varieties.

At any rate, Tavernier took many of these stones back to Europe and sold them, mostly to Louis XIV and Cardinal Mazarin. Altogether, he made six journeys to the East between 1631 and 1668. For his efforts, he was awarded the title of "official trader" for the French crown.

Tavernier is surely one of the most interesting figures of his century. Born in Paris in 1605, he was the son of a mapmaker (thus, he came by his traveling bent rather naturally) who had fled his native Antwerp in 1575 to avoid the religious persecution of Protestants. Tavernier was apprenticed to a jeweler, and his subsequent career in travel and gem dealing reflected his dual background, as he also drew some very handsome maps. He survived a shipwreck and imprisonment, and lived long enough to make a great deal of money and then lose it all. The story that he was bitten to death by a pack of dogs is unsupported, although there is likewise no evidence to suggest that he met his death any other way, either. The famous gemologist Edwin Streeter, in his 1882 *Great Gems of the World*, announced that Tavernier had met his death from "fever," which is perhaps more likely, although it is equally unsupported by any real facts. The dog-bite story may well have been concocted by the Cartiers in their attempt to add more curses and mystique to the blue stone we now know as the Hope Diamond. (For some reason, cursed stones sell better than the uncursed variety.) It is Tavernier who acquired the famous blue diamond, then known as the Tavernier Diamond or French Blue. The original French Blue was much larger than its descendant, the Hope Diamond, having been recut to improve its appearance. (In case you are wondering what makes the Hope Diamond blue, it's boron. Only about one boron atom among a million or so carbon atoms is enough to do the trick.)

Tavernier was famous for being able to adapt to foreign cultures. He learned languages with ease, and is often depicted wearing an impressive turban and fur-lined cape. The latter would have been less than useful in most of India, but it was certainly spectacular. In fact, the Indian style of dress eventually became a fashion rage in England and in Europe, where it was called Orientalism. It became particularly popular with the advance of the British Empire into India. Even Queen Victoria and her family occasionally tricked themselves out in Indian clothing and posed in *tableaux vivants*, or living portraits. Part of this was for sheer fun, but there was a serious purpose behind it as well. The Queen wanted to reduce the fear of Indians that obsessed some of her British subjects. One of her most trusted advisors, Hafiz Abdul Karim, her Hindustani language tutor and personal emissary, wore a turban, quite with the Queen's approval. Women soon adopted aigrettes and turbans for themselves, often bedecked with gems Indian style.

Vietnam

Since the 1980s, Vietnam has been producing some excellent stones originating from two geographically diverse (but mineralogically similar) sites, Luc Yen, which is north of Hanoi, and Quy Chau, which is to the south. There is a great similarity between rubies from Vietnam and Burma, not surprising since the same geographical area and conditions gave birth to both of them. The rubies that were first discovered had excellent color with a secondary pinkish hue. Recently, however, few facetable gems have been found, and the source is regarded as "unreliable." Vietnam also produces some bright pink sapphires (or pale rubies), which are quite popular.

The subjective nature of the gem trade became apparent in 1990–1991, when some Thai gem dealers bought up what they believed to be a new Burma ruby find. They went bankrupt when it was discovered that the new gems were not Burmese at all, but from Vietnam. Had the same stones really been Burmese, the speculators would undoubtedly have turned a tidy profit. It's all in the perception.

Sri Lanka

The island nation of Sri Lanka is home to the world's earliest known ruby mines, and the southern two-thirds of the island is a rich source. These rubies stem from a central mountain range where they were created and then washed down by the rivers. Rice farmers often unearth them, a pleasant addition to the rice crop. Some farmers actually lease their rice paddies to gem miners. (It's easy to get a permit to become a gem miner in Sri Lanka, and most citizens take a stab at finding gems at some point in their lives.)

Sri Lanka used to be called Ceylon, as everyone knows, but it was also once charmingly known as Serendip, a fitting name for those few who made the fortunate discovery of its gems! The lovely word "serendipity" was coined by Horace Walpole in 1754. Ultimately the name comes from the Arabic *Sarandib*, and there is a charming Persian fairy tale, *The Three Princes of Serendip*, about it. The three princes were always making wonderful discoveries, which is what the word "serendipitous" has now come to mean.

Mining began in Sri Lanka in the time of the Buddha (624–544 BCE) near Ratnapura, which is Sinhalese for "City of Gems." Sri Lanka's storied Ratnapura mine, which contains ancient stores of alluvial corundum (and, at higher elevations, stones embedded in their matrix) has been in operation since the seventh century BCE. It produces rubies, blue sapphire, and a multitude of "fancy" sapphires, including the amazing *padradascha* or "lotus" sapphire, whose delicate orange-pink hue does indeed resemble the blossom of that plant. In fact, for the ancient Sinhalese people of Sri Lanka, rubies are the Buddha's tears. The country also produces many rubies that are medium light in tone, which makes them somewhat less valuable than the best Burmese, although they have wonderful purity of color. Sri Lankan stones have both good daylight fluorescence and the right percentage of rutile; however, the fact that many are light in color can't quite make up for their excellence in other departments.

However, stones of similar color and quality have been found in the Quy Chau district of Vietnam, as well as Tanzania and Madagascar. Many experts think the name *padradascha* should be reserved for Sri Lankan stones, mostly for historical reasons. However, most authorities deem these stones sapphires and not rubies; it has furthermore been determined that in most cases their color was artificially produced by the lattice diffusion process. When they are heat-treated, some specimens can turn quite bright orange. High-quality untreated stones can sell for 50 percent more than the treated varieties. Star gems of the cherished *padradascha* shade, however, are not common; just as you seldom see a yellow or orange star sapphire. There is insufficient rutile silk in these colors to produce the star.

Sri Lanka also has star rubies and sapphires (actually, sapphires are the most common precious stones in the country) of exquisite quality. The famous Rosser Reeves Star ruby is of Sri Lankan origin. Not for nothing is Sri Lanka known as "the isle of gems." According to the seventeenth century gem trader Tavernier, rubies and the other colored stones of Ceylon came mostly from a river in the middle of the island, a river which flowed from the mountains. About three months after the spring flood, people searched the riverbanks for the precious stones. According to Tavernier, these stones were cleaner and more lovely than the Burmese ones. However, Tavernier probably came to this opinion only because he never saw the great Burmese rubies, the king having kept the best of them for himself. We also know, as mentioned before, that some Sri Lankan rubies were really from Burma.

Marco Polo (1254–1324) also wrote about Sri Lankan gems: "You must know that rubies are found in this island and in nowhere else in the world." He was wrong there, but that doesn't matter. "They find there also sapphires and topazes and amethysts, and other stone of great value," he went on eagerly. "The King of this Island owns the finest and largest in the world. I will tell you what it is like" (*Travels of Marco Polo*).

Marco Polo then went on to describe a fist-sized stone, a gem over four inches in length and as thick as his finger. According to Marco Polo, the Great Kublai Khan sent an emissary to Sri Lanka to beg for this stone, offering to pay any price for it. The unmoved king simply responded that he had received the ruby from his ancestors and would pass it on to his descendants, and that was that. What has become of this ruby, or indeed whether it even was a ruby, is anyone's guess this point. It is conceivable Marco Polo made the whole story up.

Today, mechanized mining is largely prohibited in Sri Lanka. The government believes that is best for the rubies to emerge gradually by the most natural means. This policy also keeps a large number of persons employed. At the current rate of production, it is expected that the ruby supply will last for many decades. Sri Lanka cuts many of its own gems, especially the larger ones.

India

India has a long and fabled history of trading in gems of all sorts, especially diamonds, rubies, sapphires, and pearls. At one time it was famous for its ruby production, even though it has never been the ruby center that Burma is. Nowadays, however, it produces few high-value stones, although it continues to trade vigorously in semi-precious stones and lower quality rubies. It is a major cutting center as well. India also produces a quantity of star rubies, which are in high-demand.

Karnataka in southwest India has a long history of ruby production, but tends to produce only low-grade star rubies. Orissa, in eastern India, is a newer source of stones. Relatively large crystals have been found, although they have many inclusions and are typically sold as beads or as cabochons, in which state they are prized for their magical as well as ornamental qualities.

Afghanistan

Jagdalek in Afghanistan (not far from Kabul) produces strongly fluorescent, bright-red stones rather similar to those found in Africa and Vietnam. While good information is scarce, the rubies in Afghanistan are said to be embedded in a metamorphosed marble cut by granitic intrusions from the Oligocene age and embedded in magnesium limestone deposits. Their color varies from deep red to purple. A few contain some blue. But few stones are facetable, and star rubies are not found in this region.

The most important of the ancient Afghan mines was the storied Badakhshan (also rendered *Badascian, Balakhshán, Balaxcen, Balaxia,* and *Baldassia*), located close to Shignan near the Oxus River, somewhere between the upper course of the Oxus and its tributary Turt. Badakhshan is the ultimate source of the word "balas," or low-quality red gems (including spinel). Rubies have been mined here for 6,500 years, although the product may have been more lapis lazuli than rubies.

The Badakhshan mines of Afghanistan are of ancient historical record. Marco Polo is said to have commented on the hugeness of the kingdom, the strangeness of the language, and the fact that the people there "worshipped Muhammad." However, he was basing his comments on hearsay, as he probably never actually crossed the border into the country. Still, what he has to say is interesting:

It is in this province that those fine and valuable gems the Balas [low grade rubies or spinels] Rubies are found. They are got in certain rocks among the mountains, and in the search for them the people dig great caves underground, just as is done by miners for silver. There is but one special mountain that produces them, and it is called Syghinan. The stones are dug on the king's account, and no one else dares dig in that mountain on pain of forfeiture of life as well as goods; nor may any one carry the stones out of the kingdom. But the king amasses them all, and sends them to other kings when he has tribute to render, or when he desires to offer a friendly present; and such only as he pleases he causes to be sold. Thus he acts in order to keep the Balas at a high value; for if he were to allow everybody to dig, they would extract so many that the world would be glutted with them, and they would cease to bear any value. Hence it is that he allows so few to be taken out, and is so strict in the matter. (*Travels of Marco Polo*)

And almost as anciently as Marco Polo, the Spaniard Ruy Gonzalez de Clavijo, who visited the court of Timur at Samarkand in the years 1403–1406, wrote that "Balaxia" was a "great city" ten days' journey from Samarkand towards India. Near there, it was reported, he said, that there was a mountain full of rubies, and that every day workers "broke up a rock in search of them" ("Narration of the Embassy of Ruy Gonzales de Clavijo to the Court of Timour"). It appears that the inhabitants of the area called all colored stones of the region "rubies." Sapphires were "blue rubies," yellow sapphires "yellow rubies" and so on, just as today most people call yellow corundum "yellow sapphire."

The Badakhshan mines are the source of some of the gems in major early collections, including the crown jewels of Iran, the stones of the Topkapi in Instanbul, the crown jewels of England, and the gems in the Kremlin and Diamond Fund.

Unfortunately, the historical record for these stones is a bit spotty, but no one seriously doubts their existence. The historical record also suggests that these mines were important from about 1000 to 1900 CE. After 1900, they are little mentioned.

However, the same area was the subject of an 1832 paper for the *Journal of the Asiatic Society of Bengal* by James Prinsep. He includes a report, translated into English by Raja Kalíkishen, about the original finding of the mines. The date of this momentous event is not clear. The manuscript indicated that the discovery took place in this way: It had occurred 350 years previously, but since the date of the manuscript itself is in question, this is not as helpful as it might be. The original was probably penned by Mohammed Ben Mansur, writing in the twelfth century, who stated:

The mine of this gem was not discovered until after a sudden shock of an earth-quake, in *Badakshan,* had rent asunder a mountain in that country, which exhibited to the astonished spectators a number of sparkling pink gems of the size of eggs. The women of the neighborhood thought them to possess a tingent quality, but finding they yielded no coloring matter, they threw them away. Some jewelers, discovering their worth, delivered them to the lapidaries to be worked up, but owing to their softness the workmen could not at first polish them, until they found out the method of doing so with *mark-i-shísá,* marcasite or iron pyrites...

Whether or not these stones were really rubies is questionable, as the hardness of rubies (9 on the Mohs scale) is second only to diamond, and look-alike spinels clock in at 8, which is also very hard. The stones in the report seem soft.

Captain John Wood, during an epic journey in 1872 to find the head-waters of the Oxus River, made an attempt to visit the Badakhshan ruby mines, but was forced back by bad weather. There was also a road problem. Apparently the one usable road had been destroyed by an earthquake in 1832.

In his *A Journey to the Source of the River Oxus,* he asserted the mines were 20 miles from Ish-kashm, in the Gharan district, a word itself meaning "mines." He wrote that the mines were on the right bank of the Oxus:

They face the stream, and their entrance is said to be 1,200 feet [366 m] above its level. The formation of the mountain is either red sandstone or limestone largely impregnated with magnesia. The mines are easily worked, the operation being more like digging a hole in sand, than quarrying rocks.... The galleries are described as being numerous, and running directly in from the river. The labourers are greatly incommoded by water filtering into the mine from above, and by the smoke from their lamps, for which there is no exit. Wherever a seam or whitish blotch is discov-ered, the miners set to work; and when a ruby is found it is always encased in a round nodule of considerable size. The mines have not been worked since Badakhshan fell into the hands of the Kunduz chief, who, irritated, it is supposed, at the small profit they yielded, marched the inhabitants of the district, then numbering about five hundred families, to Kunduz, and disposed of them in the slave market.

Thus, simply digging caves or fissures in limestone areas was very produc-tive and in the early days yielded some of the most splendid rubies. In addi-tion, cuttings were made into the hills where limestone croppings appeared. Water was then introduced via bamboo pipes and sluiced over the gravel.

According to an account presumably taken from a certain Mirza Nazar Báki Bég Khán, a native of Badakshán who settled at Benares, the method of mining was as follows:

Having collected a party of miners, a spot is pointed out by experienced workmen, where an adit is commenced. The aperture is cut in the rock large enough to admit a man upright: the passage is lighted at intervals by cotton *mashúls* placed in niches; as they proceed with the excavation, the rock is examined until a vein of reddish appearance is discovered, which is recognized as the matrix of the precious gem. This red colored rock or vein is called *rag-í-lál,* or, the vein of rubies; the miners set to work upon this with much art, following all its ramifications through the parent rock. The first rubies that present themselves are small, and of bad colour: these the miners called *piadehs* (foot soldiers): further on some larger and of better colour are found, which are called *sawars* (horse soldiers); the next, as they still progress in improvement, are called *amirs, bakshis,* and *vazirs,* until at last they come to the *king jewel,* after finding which, they give up working the vein: and this is always polished and presented to the king. The author proceeds to describe the finest ruby of this kind that had ever fallen under his observation. It belonged to the Oude family, and was carried off by Vizir Ali; he was afterwards employed in recovering it from the latter: it was of the size of a pigeon's egg, and the color very brilliant; weight, about two tolas; there was a flaw in it, and to hide it, the name of *Julál-ud-dín* was engraved over the part; hence the jewel was called the *lál-i-jaláli.* (Princep)

It should be noted that neither Marco Polo, Clavijo, James Princep, nor John Wood ever actually visited these mines. Everything they had to say was hearsay. The same is not true, however, of Pandit Manphúl, who wrote a report in 1867 (*Badakhshán and the Countries Around It*) firmly stating that the mines existed, although they had not been worked for the past 20 years or so. The reason was unclear to Manphúl. It was possible, he thought, that the labor was unskilled. Or perhaps the mines were simply exhausted. Or maybe the miners were just walking off with the best stones for themselves. It could have easily been a combination of factors. It is also fairly obvious that if one is supposed to hand over the largest gems to the king and keep only the smaller ones for oneself, there is a mighty temptation to take a sledgehammer to big gems. This notorious but entirely understandable practice is still another reason why large gems were quite rare.

After the turn of the twentieth century, the "new" Jagdalek mines (which lie across the border from Peshawar, Pakistan, 60 kilometers southeast of Kabul) got all the attention. These mines lie on a continental collision zone, and their rubies closely resemble the gems of Vietnam, Burma, and Sri Lanka. The rubies occur in a crystalline micaceous limestone as do those from Burma. But these mines are far from new—they have been worked for at least 700 years, and are exploited by local tribespeople using traditional small-scale mining methods. About 75 percent of the production is pink sapphire, and 15 percent is rubies. The rest is blue sapphire and mixed red/pink corundum.

Many of these stones have a pinkish or raspberry coloration, with strong fluorescence. Most of these stones are not of facetable quality (many are not very clear, owing to their thick, liquid inclusions), but those that are are said to be magnificent. (Stones from nearby Pakistan are often bright red, and a few large, over 2 inches, stones have been found there.)

One of the most noted Afghan stones is a 10.5 carat ruby that was transported to England, supposedly from the mines at Gandamak, which is about 20 miles (32 km) from Jagdalek. In actual fact, however, the stones may have been mined in Jagdalek itself, as they are very near each other.

The Irishman Valentine Ball, former head of the Geological Survey of India, also remarked on the Jagdalek mines, writing:

In the year 1879 the so-called ruby mines of the late Amir of Afghanistan, Shir Ali, which are situated near the village of Jagdalak in Kabul, were visited by Major Stewart of the Guides. Two specimens of stones, called *yakut* by the natives, and samples of the matrix, were forwarded to the office of the Geological Survey for examination. The stones were determined to be "the best of their kinds."

Based partly on this evaluation, it has been suggested that the stones under consideration were spinels. Still, because we have confirmed recent ruby finds from the area and because rubies come in various qualities, we cannot simply dismiss the idea that they were indeed rubies, as the Amir of the country Shir Ali had declared. He apparently kept a strict guard over the mines and only allowed special friends to work them.

The area where these mines were presumably located is now in Tajikistan, right on the border with Afghanistan. Curiously, in the late 1980s, large spinels of a reddish tint were reported from the Pamir Mountain region of that country. Possibly they hail from the mysterious mines of Badakhshan. The vagaries of the political situation may keep us from ever knowing for certain.

During the 1980s, when the Russians occupied the country, all Afghan gem and mineral mining was controlled by the state, at least nominally. However, many mines lay in inaccessible areas, and their products became important source of income for the rebels.

Australia

Australia is not world-renowned for its rubies, but it has a few. A few small stones are mined in pegmatites in the New England region of New South Wales. These stones occur in the gold sands of the Cudgegong River and its tributaries near the town of Mudgee (populaton 7440). Mudgee is also known for its wine, unbelievable as that may sound.

In 1999 rubies were found at a farm in Gloucester (also in New South Wales) in an ancient riverbed. The riverbed is expected eventually to yield millions of carats of facetable stone. This wasn't a run-of-the-mill farm, though. It was one owned by billionaire Kerry Packer, who used it mainly for polo pony practice. It seems unfair, but that's the way it is. The stones are similar to Thai ruby, with quality ranging from poor to very good. Actually, Australia is more famous for its red garnets—often passed off as "Adelaide Rubies."

Africa

Many believe that Africa is going to be the world's next best source for new ruby finds, although evidence for this optimistic assessment is scanty at this time. Most African stones are heavily flawed. Like other rubies around the world, African stones are shipped to Thailand for cutting. Kenya and Tanzania produce sheet-like, purplish-red gem material with good fluorescence. Some Thai dealers buy these rubies hoping to find gem material that is good enough to facet. It rarely happens, but when it does the stones are rated of excellent quality.

The Longido mine in northern Tanzania is famous for very large, opaque ruby crystals surrounded by green zoisite, which produces a distinct "watermelon effect" to the stone. These rubies were first discovered in 1949 by Tom Blevins, an English prospector who had lived in nearby Kenya for many years. Blevins has spent many discouraging months on the hunt for gems, and was close to giving up. Then late one afternoon, he came upon a small flat sink or basin, devoid of plant life, that extended for several yards to the base of an outcrop of zoisite. What he saw took his breath away: an amazing number of sharp-faced, hexagonal, deep red ruby crystals. Some measured two inches across and weighed hundreds of carats. So intoxicated was he by the sight, as he began picking up loose crystals by the handful and stuffing them into his duffel bag, that the idea flashed across his mind that he had found the storied mines of King Solomon. Sadly, closer examination later showed that the best, clearest, most deeply colored stones occurred only at one end of the outcrop. These high-quality stones were very small—the larger crystals invariably were shown to be of inferior quality. Longido would not turn out to be another Mogok, although the largest crystal weighed in at slightly more than 30,000 carats. From the standpoint of quantity, Longido would prove to be one of the top ruby producers of all time.

However, things did not proceed as smoothly as one might hope. There was, for instance, some opposition from the local elephant herd. Apparently

African elephant are more protective of their mineral rights than Burmese ones. Day after day, the miner would return to the mining area only to find the road blocked off by elephant-toppled trees. This occurred on such a regular basis that it soon became clear the animals wanted the miners to clear out, either because they simply disliked human trespassers or because they wished to have the stones for themselves. They would not, after all, be the first elephants to be decked in rubies.

The best African mine is probably Kenya's Aqua Mine, located north of the Tanzanian border. Another mine, called the John Saul Mine, produced excellent stones on a par with those from Burma; however, not many are mined there nowadays.

Gem traders are in the habit of calling Kenya's line of ruby mines, which run east to west near the Tanzanian border, "Penny Lane"; they are better known for their quantity of rubies than their quality.

Madagascar also produces some rubies (two important ruby deposits have recently been discovered there), and there are some in Malawi as well. Most of the best is cabochon quality. The stones are very red, but they have a dull appearance. The rubies in Madagascar set off quite a "ruby rush," with between 20,000 and 30,000 miners descending upon a quiet rural village. The gem bearing area is a 15-hour trip from Andilamena by four-wheel drive, followed by 30 miles on foot. There are no real accommodations to be had, and the enterprise is considered very dangerous, not least because of the large amounts of cash involved. In fact, the area is so remote that the government has no way to control what is happening there.

6

The King of Stones—and the Stone of Kings

From the dawn of history the great and the powerful have collected the red stone. Cleopatra liked them (although emeralds were her favorite), as did Messalina, the promiscuous third wife of the emperor Claudius, and so did Mary Stuart. Cardinal Richelieu and Marie de Medici had famous collections of them. And while their source is usually Burma, most came into their fame and glory in other lands.

RUBIES COME TO INDIA

One very famous ruby is the Chhatrapati Manik, an oval Burmese cabochon of superior color. Its history is supposed to go back 2,000 years to the reign of Sri Raja Bir Vikramaditya, King of Ujjain (in present day Madhya Pradesh, India). He gave himself the title "Chhatrapati," which means "Supreme King," and upon the advice of his court astrologers placed a great ruby in the crown he commissioned for himself. The crown contained nine major gems, according to the Jyotish system of astrology popular in India. Nine (three times three) is considered the most powerful of all numbers, and the ruby is associated with number nine.

Each stone in this *navaratna*, as it is known, represents a heavenly body, and the ruby is the sun. There is significant magical power associated with the navaratna, as we shall see, and the more perfect the stone the greater the power. It wasn't easy finding the perfect stone, but Bir Vikramaditya, insisted and eventually a deep red Burmese ruby (weighing between 20 and 40 carats, depending upon the source) cut in cabochon was located in the collection of a banker named Chhatrapati, after the king himself. However, after many years, the crown was traded away to the ruler of Golconda, Sultan Abul Hassan Qutub, or Tana Shah (1672–1687), who destroyed the crown after ripping out the gems from it. The ruby was his favorite, and the shah had his name engraved upon it and even commissioned a book of poetry to be written that glorified the stone.

When the Mogul Emperor Aurangzeb defeated Tana Shah in October, 1687, after a siege of many months, he is said to have taken Tana Shah prisoner (the sultan later died in captivity). Aurangzeb's own son led the battle and brought not only the ruby, but the book of poetry celebrating it, to his father. Aurangzeb had the former ruler's name scratched out and his own substituted for it. The history of this stone at the point seems to become confounded with that of the Timur Ruby (which is not a ruby), upon which Aurangzeb also had his name recorded.

At a later time Aurangzeb gave the stone (and the poetry book) to some bankers in Murshidabad, Bengal, in recognition of their services to the Mogul Empire. The bankers were supposedly already the richest in the world, and as usual, it is the rich who end up with the best presents.

After that, a certain Lala Kalkadas from Lucknow traded a bunch of jewels for both the ruby and the book, and forthwith ground off Aurangzeb's name. Unfortunately the book was lost during the Indian Mutiny (1857–1858), but Lala Budreedas, son of Lala Kalkadas, hung onto the stone for dear life. He moved to Calcutta, and had the stone mounted in a brand new tiara. Then it disappeared, although it was said to have been sold and subsequently made an appearance in London. As you might expect, no one knows where it is now.

RUBIES COME TO PERSIA

The Central Bank of Iran in Tehran contains the finest Crown Jewels in the world, making the British crown jewels seem rather pitiful in comparison. This seems a bit strange, since Iran has no gem resources of its own. (However, the same can be said for England.) Some may think the Iranian jewels gaudy and over the top, but if one cannot overindulge one's taste in the area of crown jewels, one cannot do it anywhere. Some of the jewels are emeralds acquired from the Spanish in the sixteenth century (who stole them from the Colombians). Most of the rest were stolen from India in 1739, when Nadir Shah conquered Delhi and returned home to Iran with everything he could get his hands on, including the fabled Peacock throne. (There was actually more than one Peacock throne, and the original throne was broken up, although many of the gems that adorned the throne are still in existence.) Perhaps the crown of the Iranian Crown Jewels is its collection of Burmese rubies, including some wonderful 8–16 carat cabochon stones. Before the fine art of faceting was well understood, most stones were cut in cabochon, as star rubies must be. The collection also boasts some record-breaking red spinels.

The Iranian Crown jewels also include a massive belt buckle adorned with 20 matched and matchless cabochon rubies, some of which weigh 10 carats. After the 1978–79 Revolution, the collection was put away. Some Islamic fanatics insisted that the jewels be sold off, but luckily wiser heads prevailed, and the collection was kept intact, and quietly re-opened for public viewing in February 1992.

The Iranians developed an intriguing way of keeping track of their best stones. Rather than just have them scattered all over the palace or crammed in a treasure chest, an early Persian ruler, Nasseridin Shah (1848–1896), lit upon a unique idea, a sort of inventory revolution. He simply had thousands of his loose gems made up into a stunning globe 45 centimeters in diameter and 110 centimeters high. Here the continents were picked out in rubies or spinels and the oceans in Colombian emeralds (one of which weighs in at 175 carats). The biggest ruby is 75 carats (massive for a ruby) and the largest spinel is 110 carats. He saved the diamonds for accents. In this way, it would have been immediately and picturesquely clear when a gem was taken. The method seemed to work, as all the countries remain intact even today. This perhaps overly impressive artifact contains 51,000 gemstones and 34 kilograms of pure gold.

RUBIES COME TO ENGLAND

Since England ruled Burma for more than half a century, it would be strange indeed if some Burmese rubies didn't make their way to the British. In fact, at one time, the gold coronation ring of the British monarchs bore a very large table-cut ruby, upon which was engraved the cross of St. George, patron saint of England. Henry III (1216–1272) inaugurated the custom of using a ruby as the central stone in the ring, although George IV (1920–1830) replaced it with a sapphire. Rubies were also traditional for the Queen Consort ring. There was a bit of a kerfuffle in this regard in 1689 during the joint investiture of William and Mary, when Mary's coronation ring was put on the finger of William by mistake. This should have been expected; it is difficult enough to stage the coronation of one monarch; a double coronation is asking for a lot. In 1831 Queen Adelaide, wife of William IV, had another ruby coronation ring made, a 50.15 carat stone that is currently part of the crown jewels.

Prince Andrew presented Sarah Ferguson with an oval pigeon blood ruby (worth $37,000) and diamond ring when they got in engaged in 1986. The ring was crafted by Garrard's, the crown jeweler and the same people who made Princess Diana's sapphire engagement ring. The central ruby was flanked by 10 drop diamonds. Prince Andrew had asked specifically for a ruby in the belief that it would go with Fergie's red hair. He also helped design the ring. Plain diamond rings have never been as fashionable in England as in the United States. Of course, Andrew and Sarah are divorced now, but that is beside the point. Jackie Onassis was given a ruby cabochon engagement ring "the size of an Easter Egg." It reportedly sold for $2.6 million after her death.

The list of England's great rubies is an elite one, and it's difficult to say which among them is the "best." As with many great stones, the history of their travels is fascinating in itself.

One of the most famous is the Edwardes Ruby (167 carats), which was donated to the British Natural History Museum in 1887 by the writer and art critic John Ruskin, along with the Colenso Diamond, a perfect 133 carat octahedron. Although it is not widely known, Ruskin had an abiding interest in mineralogy all his life. His work *Ethics of the Dust* appeared out of a series of lectures given at Winnington Hall Girls' School in Cheshire, in which he constructed his lessons around crystals.

This gem was named in honor of Major-General Sir Herbert Benjamin Edwardes (1819–1868), who "saved" British rule in India during the time of the Indian Mutiny of 1857. Like many of his era, Edwardes basically learned his trade while working for the East India Company, eventually acquiring a distinguished military and diplomatic record. He was severely wounded at the battle of Mudki, and later had his right hand blown off when a pistol in his belt exploded. He was considered quite a hero in England, which is presumably why Ruskin honored him by naming the ruby after him. (Apparently there was some friction concerning the naming of the stone, as revealed in Ruskins's letters.) The stone is a detached, deep red, squat crystal, still rough. Although the upper portion is transparent, the bottom part is quite flawed. It is probably from Burma, although this is not certain.

The same Natural History Museum is also home to the Pain Collection, donated in 1971, which consists almost entirely of stones (mostly cut) gathered from the Mogok area. The star of this collection is perhaps a very large uncut single ruby. It shows deep etching, presenting a rather "fretted" appearance, according to Peter Tandy, Curator of Minerals at the Natural History Museum. Of more sentimental value is a small marquise-shaped ruby that belonged to Sir Hans Sloane (1660–1753), a physician born in Ireland whose collections of 50,000 books and 3,560 manuscripts formed the nucleus of the British Museum. Since the Natural History Museum is an "offspring" of the latter, Sloane may properly be regarded as the spiritual founder of both institutions. Even more important to some, Sloane introduced cocoa to Great Britain after first trying it out himself while in Jamaica. He found it "nauseous," but he thought he could improve its palatability by adding milk. For Sloane, the best use for cocoa was as a medicine, and he duly added it to his pharmacopoeia.

The British Museum didn't get every good ruby that showed up in England, however. In 1875, for example, the Burmese royal family sent two large rubies to England for sale. (They were not interested in museum collections.) The rumor was that the cash-strapped king, Mindon Min, was forced into their sale, but this is not proved. It is also surmised that he kept back stones of even higher quality, but this is not proved either. The stones were considered to be of such superb quality that they were escorted by a military guard on the way to the transport ship. One gem weighing 37 carats was cushion shaped; the other, in a teardrop shape, weighed 47 carats. They were recut by James N. Forster to 32 5/16 and 39 9/16 carats (or 38 9/16 carat, depending on the source) and sold for 10,000 and 20,000 pounds respectively. They are called the J.N. Forster rubies, but their present location is uncertain.

The "Mandalay Ruby," a 48.019-carat cushion-shaped ruby, was offered for sale by Sotheby's New York on October 18, 1988. However, it received no bids, even though Sotheby's hinted that perhaps it was the lost 47-carat J.N. Forster Ruby. In once sense, this isn't very likely, since rubies don't generally expand, although since the weight of the carat was not standardized, it is conceivable. However, it would have had to expand considerably, since, as we have seen, Forster recut 47 carats down to 39 carats and change. Quickly re-treading, Sotheby's suggested obliquely that it might be another famous although unnamed ruby weighing 46¾ carats; oblong in form (probably cut), mounted in a brooch with four brilliant-cut diamonds and sold (or bought in) at Christie's of London on May 7, 1896 for 8,000 pounds. This is equally unlikely, as the stone Sotheby's offered for sale was definitely not oblong. At any rate, the stone was withdrawn from the sale, and there is no telling where it is now.

In case you're wondering about those most famous stones, the Black Prince Ruby and the Timur Ruby, we'll get to them later. It turns out they aren't rubies.

RUBIES COME TO AMERICA

The United States has its own share of fine rubies, most coming from Burma, but some homegrown. One of the most notable rubies is the massive 196.10 carat Hixon Ruby of the Los Angeles County Museum of Natural History, donated in 1978 by Frederick C. Hixon. This is one of the best and most perfect Burmese gems on public display anywhere in the world. It looks vaguely like an arrowhead (at least from certain angles), and was found in marble rock.

The 100.32-carat Edith Haggin de Long Ruby resides at the American Museum of Natural History in New York. This is another great star ruby, mined in Burma during the early part of the twentieth century. It was sold by to Edith Haggin de Long by Martin Ehrmann for a mere $21,400. In 1937 she donated the stone to the Museum. This ruby was one of those stolen by the infamous Jack Murphy and two accomplices from the Museum in the biggest jewel heist in American history, the so-called Great Jewel Robbery. Murphy was a rather unusual combination of murderer, surfing champion, and jewel thief. (He also claims to have played violin with the Pittsburgh Symphony orchestra, and to have been a tennis pro and movie stuntman, but that is another issue.)

On October 29, 1964, he snatched the Star of India (the world's largest star sapphire), the Eagle Diamond, and the de Long Ruby. At the time, the stones were valued at more than $400,000. The thieves had cleverly unlocked a bathroom window during a visit to the museum during regular hours, and climbed back in at night. To their delight, they found that the only gem in the entire collection that was protected by an alarm was the Star of India—and the alarm battery was dead. (The Museum had not even bothered to insure the stone.) One of the stories about rubies is that they grant the gift of invisibility to thieves; perhaps this was an additional factor in the success of the heist.

After payment of a mere $25,000 ransom (negotiations took 10 months), it was retrieved at the designated drop off site, a phone booth in Florida. "Murph the Surf," as he was agreeably known, was arrested two days later with his accomplices. The Star of India was in a locker in a Miami bus station. Most of the other gems were also recovered, except the Eagle Diamond, a 16.25 carat stone actually discovered in Wisconsin in 1876 by a man digging a well. At the time it was one of the largest diamonds ever discovered in the United States. It has never been found after the heist and was probably cut up into little engagement rings.

The thieves received, rather unbelievably, a mere three year sentence. After getting out of jail, Murphy started killing people. In 1968 he was convicted of first-degree murder of a California secretary, one of two women whose bodies were found in Whiskey Creek near Hollywood, California, the year before. He was sentenced to life in prison, where he learned to paint seascapes and lighthouses. He also, he claimed, found God. He was paroled in 1986. Murphy embarked on a career as an evangelical preacher and is now the director of an international prison minsitry. In 1975 a movie was made about his career to date, uninspiringly named *Murph the Surf.*

The New York Museum of Natural History also holds the 116.75-carat deep purplish-red Midnight Star Ruby, which as its name suggests is indeed a star ruby. It was mined in Sri Lanka, and to date has not been stolen.

The 138.7-carat Rosser Reeves Ruby is in the Smithsonian Institution in Washington, D.C. This is probably the largest and finest star gem in existence, better than the de Long, with more translucency and a better star. Unfortunately, nothing is known about its history. It appeared rather suddenly in London in the late 1950s, at that time weighing in at 140 carats, but was later re-cut to "center" its star. The stone was named for its donor, and in 1966 insured for $150,000. What makes this stone also unusual among the world's finest rubies is that it comes not from Mogok in Burma, but from the gem gravels of Sri Lanka, which are famous for their fabulous star stones.

The Smithsonian's National Museum of Natural History also fairly recently came into possession of one of the world's largest and finest faceted ruby gemstones, an oval-shaped 23.1-carat Burmese ruby. The ruby was donated by businessman, nuclear physicist, and philanthropist Peter Buck in memory of his wife Carmen Lúcia, who died in 2003. (Buck is also known for helping a friend buy a sandwich shop that later became known as Subway.) Carmen Lúcia was a collector and researcher in her own right. She and her husband tried desperately to acquire the stone when they heard it was coming on the market (after many years in private hands). Although she died from cancer before they were able to complete the purchase, Buck made sufficient funds available to the Smithsonian for its purchase.

This spectacular stone, mined from the Mogok region in the 1930s, combines a rich, saturated color with extraordinary clarity, rich red with slight pink and purple undertones. In addition to its extraordinary size for a cut gem, it was the fortunate recipient of an elegant cut that offers vivid color reflections. The stone is set in a platinum ring and flanked by two triangular-shaped diamonds. The ruby is now is on display where it can be seen by the general public, and which indeed now belongs to the American public.

The Appalachian Star ruby, a 139.43 carat stone, was found by Jarvis Wayne Messer in his native North Carolina, along with the "Smoky Mountain Two Star Ruby," an 86.56 carat round double Star Ruby, which displays a perfect six-pointed star. It is considered to be the world's heaviest ruby, and went on public display at the Natural History Museum in London in 1992. This is just slightly heavier than the Rosser Reeves Ruby from Sri Lanka. Both stones were the subject of a subsequent lawsuit in October 2005 to settle some debts.

Kings and queens aren't the only people to be drenched in rubies. So was Elizabeth Taylor—almost literally. In her autobiographical book, *My Love Affair With Jewelry,* she writes about the time husband Mike Todd presented her with an astonishing Cartier ruby-and-diamond necklace. "I was in the pool, swimming laps at our home, and Mike came outside to keep me company. I got out of the pool and put my arms around him, and he said, 'Wait a minute, don't joggle your tiara.'"

The Queen of England might not wear a tiara while swimming laps, but apparently this was ordinary practice for Taylor, who wore the one Todd had given her, telling her that she was his "queen." She wore it for the first time at the Academy Awards, when Todd's film *Around the World in 80 Days* won for Best Picture. However, she was also sensible enough not to wear her tiara when she met the real Queen Elizabeth in 1976, as the Queen was wearing *hers.* It doesn't do to have competing tiaras.

At any rate, in reference to the ruby and diamond necklace, Taylor continues breathlessly,

He was holding a red leather box, and inside was a ruby necklace, which glittered in the warm light. It was like the sun, lit up and made of red fire. First Mike put it around my neck and smiled. Then he bent down and put matching earrings on me. Next came the bracelet. Since there was no mirror around, I had to look into the water. The jewelry was so glorious, rippling red on blue like a painting. I just shrieked with joy, put my arms around Mike's neck, and pulled him into the water after me.

The whole episode was captured on camera in a home movie taken by Eve Johnson, wife of actor Van Johnson.

Taylor also owns a fabulous ruby ring, given to her by another husband, Richard Burton. This one was a Christmas gift, shoved in the bottom of a Christmas stocking. It was designed by Van Cleef & Arpels and "perfect," according to Taylor. Burton said that it had taken him four years to find that stone. Rubies were his own favorite, "red for Wales," as he liked to say.

Nor was our former First Lady Jackie Kennedy Onassis a slouch in the jewelry department. She owned a fabulous 17.68 carat ruby ring (sold at auction at Sotheby's in April 1996 for $290,000), a pair of cabochon dangling ruby earrings ($360,000), and a cabochon ruby necklace ($247,500). She also had a cabochon engagement ring the shape of an Easter egg that sold for $2.6 million after her death. That is nothing compared to a ruby bracelet once owned by Marlene Dietrich and worn by her in the film *Stage Fright*—it went for $990,000 at Sotheby's.

RUBIES COME TO FRANCE AND BEYOND

France has also had its collection of great rubies. One of the most spectacular of the French crown jewels began life as a ruby and diamond parure that Napoleon ordered in 1810 for his new wife, Archduchess Marie-Louise of Habsburg. A parure is a set of matching jewelry, in this case consisting of a coronet, diadem, comb, earrings, necklace, belt, and pair of bracelets. Marie-Louise was the daughter of Emperor Franz I of Austria and niece of Marie-Antoinette. (In case you're wondering about Josephine, Napoleon had divorced her the year before.) The designer was Franois-Regnault Nitot. Another parure was made of turquoise and diamonds. Napoleon was extremely fond of turquoise. Both parures were delivered to Napoleon on January 16, 1811.

However, after Napoleon lost power, the House of Bourbon (represented by Louis XVIII) resumed both the crown and the crown jewels, including the ruby-and-diamond parure, which thus became part of the crown collection. As Louis XVIII was a widower, the parure was worn by his niece Marie-Therese, the daughter of the decapitated Louis XVI and Marie-Antoinette.

In 1816, the entire suite was redesigned into a much simpler affair with fewer stones, resembling something much closer to today's taste. The new design was sketched out by Evrard Bapst and actually made by Bapst's father-in-law, Paul-Nicholas Mnire, who had been sort of a royal jeweler back in the days of Louis XVI. In most cases, the original stones were re-set without being re-cut.

Apparently Bapst inherited his father-in-law's position as jeweler to the crown, for in 1825, he redesigned the parure yet once more—this time for the coronation of Charles X. He added some more rubies, apparently feeling that the earlier, simpler design needed to be augmented for such a momentous occasion. Charles didn't wear them, of course; the honors were done again by Marie-Therese, who was Charles's daughter-in-law. The whole set continued to evolve, and by the end of Charles's reign, it consisted of a diadem, a coronet, a belt, a pair of bracelets, earrings, a pendant, 14 corsage studs, a rosette-shaped fastener, two "accessories," a "small necklace," and the "big necklace." Things were getting rather out of control. The French government changed hands again, reverting to Napoleon III (1808–1873), who became emperor in 1852. Before that he was only the "President," but he thought the title of Emperor was more suitable. The ruby suite was passed along to his wife, the Empress Eugenie. For once it wasn't torn apart and redesigned, as the Empress thought everything was perfectly fine the way it was. The suite was one of the only crown jewels to remain intact, however; most of the rest of the collection was done over again.

Empress Eugenie was the last member of the royal family ever to wear the magnificent parure. In 1887, the government auctioned it off, along with the rest of the crown jewels. The whereabouts of only about half the pieces are presently known. Some of the most magnificent of the ornaments were dismantled, with the large diamonds sold off individually. The diamond-and-ruby parures were also split up and sold one by one. (The largest part of the parure was the belt, when it was broken up and sold in ten lots.) The fate of most of the jewels remains unclear, although a few of them are in the Louvre. The "large necklace" was bought at an 1887 auction by Bapst and Son, the same company that had done its remodeling. It vanished from the scene for many years, not re-appearing until it was sold at Christie's Geneva auctions in 1982 ($458,000) and again in 1993 (nearly $1.3 million). Gem historian Mary Hammid strongly feels that the small necklace may have survived in a slightly re-arranged state to be offered at auction in 1980 at Christie's, along with a pair of earrings. (However, since the bidding did not reach the confidential minimum of about $60,000, the set went unsold.)

The diadem was sold at the 1887 auction to a certain "Mr. Hass," and the matching bracelets went to Tiffany, who resold them to Mr. Bradley Martin (1841–1913), a charter member of New York's "Four Hundred" club. He and his wife Cornelia (1845–1920) combined their last names after they returned from England in 1892, apparently deciding that this adventure in nomenclature gave them more snob appeal. (Bradley Martin's brother, Frederick Townsend Martin, is perhaps best known for saying, "We are rich. We own America. We got it, God knows how, but we intend to keep it.")

Eventually, the diadem ended up on the person of the aforementioned Cornelia Bradley Martin at a controversial Costume Ball of 1897. This event was originally devised as a charity fundraiser for the thousands of people ruined in the economic panic of 1896–1897. Mrs. Bradley Martin thought that the florists, decorators, boot-makers, wig-makers, hair stylists, milliners, and seamstresses the party would require would bring in enough revenue to bring the city to rights again. To this end, she sent out invitations at rather the last minute, so that people wouldn't have time to order their costumes overseas and would have to rely on local help. The party was scheduled for February 10, 1897, and the theme was "the Court of Versailles at the time of Louis XIV." It was to be held at the Grand Ballroom of the Waldorf Astoria, rigged up to look like the Great Hall of Mirrors at Versailles. As is customary with high society, the ball did not begin until 11 PM, when all the really poor people were in bed resting up for another day's hard work.

Mr. Bradley Martin was dressed up like Louis XV, which rather belies the theme, but Mrs. Bradley Martin was even more incongruous, dressing up like Mary Queen of Scots, who had been dead for about 200 years by the time of Louis XIV. However, she did wear those ruby-and-diamond bracelets, joined together to make a choker, as well as lots of other jewelry. In fact, she wore about as much jewelry as one person *can* wear: a diamond tiara, a diamond and ruby necklace, a diamond sunburst brooch, another diamond brooch, a diamond studded belt, and bands of diamonds draped from her shoulder to her waist like a bandolier of bullets. She also sported some pearls that had belonged to Josephine, wife of Napoleon. She stood in front of a set of rare tapestries and under a canopy of velvet to welcome her 600 guests, one of whom showed up as Pocahontas.

The *New York Times* rather cattily suggested that not nearly as many guests attended as were expected, and that quite a number of those who showed up left early. "Fully half the carriages were ordered before the time set for the cotillion," remarked the writer rather smugly, and of those who stayed, many seemed content to "wander about the hotel corridors." The *Times* also offered a gratuitous anthropological commentary, saying that

> The Latin races can scarcely imagine any entertainment of note that is not accompanied by fancy dress, and the carnival spirit pervades all their diversions. Northern peoples, while not as much addicted to the arraying in costumes of other peoples, times, and lands, still feel impelled at intervals by a curious sense of personal vanity to impersonate characters other than their own, and to strut up and down in apparel that present fashions and their more practical life could never countenance except as a diversion.

It may not have been a social triumph, but the New York City tax board immediately doubled the Bradley Martins' tax assessment, a decision based largely on the amount of jewelry worn at the infamous costume ball. *Colliers* and other magazines had a heyday of poking fun at the rich people, too. Soon, the Bradley Martins had had enough and moved back to England, which they always considered their real home anyway.

In 1893, at the age of 16, Cornelia had managed to capture (in wedlock) William George Robert (1868–1921), the Fourth Earl of Craven, and thus became a very young "Lady Craven." There was considerable talk about who married whom for money and prestige, although we don't need to go into that. The Earl did have 40,000 acres and three estates, but Cornelia helped out with the payments. In 1921, this Earl fell off his yacht while racing at Cowes Week and drowned. There was some sort of curse afloat that whoever inherited the Earl's title would die young, although the Earl wasn't all that young.

The curse did not extend to Lady Craven, however, who hung on until 1961 along with most of the jewelry she inherited form her mother, including those diamond-and-ruby bracelets. They were sold at Sotheby's in November 1961. The bracelets were bought by a London jewelry firm, which then sold them to Claude Menier, who bequeathed them to the Louvre. And that's where they are today.

One of the most famous French pieces of jewelry ever created was the special Insignia for the Order of the Golden Fleece, commissioned by Louis XV of France (1710–1774). It all happened because Louis XV became a knight of this order in 1745. (He probably chose this order because he was a direct descendant of Philip the Good, who founded the order in 1429 or 1430. It was Philip the Good and his Golden Fleece knights who sold Joan of Arc to the English, who had her killed.)

The eighteenth century was a little late for knighthood in the true sense of the word, but it was a romantic idea, nonetheless. Every order of knights had its own particular insignia, which all the knights wore when they gathered together for their knightly activities. By the eighteenth century (and even before), none of this amounted to much more than a modern day Moose meeting, but it probably kept the men out of trouble.

The regular Insignia for the Order of the Golden Fleece was a chain with a golden ornament that supposedly represented the fleece of a sheep. (The name of the order comes form the Greek myth about Jason and the Golden Fleece, of course.)

Knights could own the standard package, or they could order extras. Louis XV wanted extras. To that end he commissioned his court jeweler, André Jacquemin, to design something really special. Jaquemin ordered up a few hundred gems (rubies, sapphires, topazes, and diamonds) and had his lapidary Jules Guay do all the cutting work. It took two years, but the resulting piece, called the *Toison d'or de la parurer de couleur* ("the colored version of the Golden Fleece") to distinguish it from the ordinary or "white" version, was a tremendous hit.

The piece featured two great stones: the French Blue (the ancestor of the Hope Diamond) and the Anne of Brittany "Ruby," which is also called the Côte de Bretagne Jewel, a 105 carat polished but irregularly shaped orange-red spinel. This stone is considered by some to have been a part of Charlemagne's Imperial Crown (now vanished). The crown also may have included, according to Eduard Gübelin and Franz-Xaver Erni, a massive (241 carats) spinel known charmingly as "the egg of Naples." (The French have always been rather unsentimental about their crowns, never hesitating to pawn them, melt them down, or sell off their constituent elements in times of national emergency.)

Anne of Brittany (1477–1514) was Queen of France from 1491 to 1498 as a consort of Charles VIII. She then took up with the divorced Louis XII from 1499 until her death. She wore white for her wedding, the first woman on record to have done so and beginning an enduring tradition, even though it was her third marriage. (This popular woman had first been married to Maximilian I by proxy but the marriage was annulled.)

Even though she had so many offers of marriage, she was not exactly a femme fatale, having one leg shorter than the other, a not uncommon birth defect. But she was brilliant, and that always counts for something. Anne was also quite fond of jewelry of all sorts and kept a box of gems. Whenever she was pleased by a visitor, she would randomly pick up a stone and hand it over to the guest. She was a popular hostess.

Anne's Côte de Bretagne spinel formed the head and body of the dragon in the Insignia; the tail was made of 200 small diamonds curled around a 42-carat brilliant cut diamond. The outstretched dragon wings were made of diamonds too. The dragon's mouth issued golden flames sparked by topazes, which encircled the French Blue.

The whole ornament was about five inches long, and the king liked to wear it suspended from a red ribbon attached to a 24 carat hexagonal diamond. When the French Crown Jewels were inventoried in 1774, the value of the Insignia was estimated at 7.3 million dollars (in today's money, of course. The French didn't have any dollars, they had *livres*, which were not the same thing at all.) Richard Kurin, in his fascinating account of the French Blue, *The Hope Diamond: The Legendary History of a Cursed Gem*, wrote that

why Louis XV would have desired such an elaborate insignia of the Order is not explicitly known; one reason might be that he was a direct lineal descendant of

Phillip III, known as Philip the Good, Duke of Burgundy, who founded the order of the Golden Fleece in 1430. Another might be that he found the story of the Order fascinating and what it stood for consistent with his own perspective on pre-Revolutionary France and Europe.

My own suggestion is that he ordered it because he had a lot of money and rather execrable taste. Even his furniture is not to everyone's liking. Cabriole legs are not all they are cracked up to be.

No other reason is really needed. The Insignia was passed on to Louis XVI, whose taste was even more depraved than that of his predecessor. We all know what happened to him. The Insignia was stolen from the royal warehouse during the French Revolution along with most of the other Crown Jewels. It was broken up, and its constituent jewels underwent their separate fates. The spinel dragon showed up years later in Germany, disappeared again, and somewhat mysteriously was returned to France. It is now presumably safe in the Apollo Gallery, in the Louvre Museum.

The Côte de Bretagne spinel and what remains of the French Blue (in other words, the Hope Diamond) had a brief reunion in the 1960s. In 1962, Pierre Verlet, the head curator of the Louvre, requested the loan of the Hope for a month in May during an exhibition of the historic jewels of France, "Ten Centuries of French Jewels." His request was turned down, with the Smithsonian worried about security, claiming that May was their busy month, and so on. Verlet hinted that France might be able to reciprocate in the future with the loan of something equally notable. Again he was refused. However, the French had friends in high places—namely First Lady Jackie Kennedy, who twisted enough arms in the nicest possible way to make it happen. And the French did reciprocate the following year—with the Mona Lisa.

RUBIES COME TO LUXEMBOURG AND THE CZECH REPUBLIC

The largest gem quality ruby in existence is presumed to be a 250 carat gem (measuring 1.55 by 0.55 inches). It is part of the coronation crown of Charles IV of Luxembourg (1316–1378). His original name was Wenceslaus, but he changed it to something more pronounceable at his confirmation. Perhaps Charles née Wenceslaus felt guilty for changing his own name, for he commanded that his crown (ordered in 1346) was to be permanently deposited in St. Vitus Cathedral (located in the Prague Castle) in which was a shrine containing the skull of his erstwhile namesake St. Wenceslas (c. 903–935), Duke of Bohemia and patron saint of the country. (He was murdered by his brother Boleslaw. Wenceslas's body was hacked to pieces, but Boleslaw repented of his deed three years later and ordered the pieces gathered together. Yes, this is indeed the "good king Wenceslas" of Christmas carol fame, although he wasn't a king and the events in the song are mythical.) More specifically, Charles wanted the crown stashed on "the saint's head," really only the golden head of the Saint's bust, which was kept along with his remains (apparently atop the skull itself) in a dark chamber of the Cathedral.

For the rest of his life, Charles kept tinkering with his crown, much like Thomas Jefferson and Monticello, adding first one gem and then another, until it reached its present form. He ordered that the crown remain unchanged after his death, apparently deciding that it had reached perfection. This was a wonderful gift to future gemologists, who have had the pleasure of examining the stones and acquiring a great deal of knowledge about their provenance. Currently, it is believed that most of the gems in the crown came from southern Asia. The crown, wrought of extremely pure gold (21–22 carat), also contains 19 sapphires, 44 spinels, 30 emeralds, and 20 pearls, many of which are extremely large. It weighs two and half kilos (about five and a half pounds).

However, the crown did not remain at St. Vitus. Instead it led a peripatetic life in response to political unrest and threat of war. It was always trotted out for coronations, however, until monarchy ended, and now resides with the rest of the Czech crown jewels, once again locked securely in St. Vitus Cathedral. The President of the Republic has the exclusive right to decide when the Crown Jewels are displayed, but the process is difficult. The door to the chamber and the iron safe where the jewels are kept have seven locks, and there are seven holders of the keys: the President of the Republic, the Prime Minister, the Prague Archbishop, the Chairman of the House of Deputies, the Chairman of the Senate, the Dean of the Metropolitan Chapter of St. Vitus Cathedral, and the Lord Mayor of Prague, who must all convene and participate in the opening of the door and the safe.

Legend tells us that anyone who wears this crown without lawful right will die within a year. Certainly, Reinhard Heydrich, the Nazi "Protector" of Czechoslovakia, whose pleasant nicknames were the Butcher of Prague, the Blond Beast, and the Hangman, met a swift end at the hands of British-trained Czech partisans in 1942, less than a year after he took control. It is a pleasure to report that the Heydrich died in agony from blood poisoning. The grenade fragments apparently drove contaminated bits of horsehair upholstery from Heydrich's car into his wounds. Hitler retaliated by having all men over the age of 16 in Lidice murdered and another nearby village burned. It is a wonder that he didn't have the horses killed as well for their part in having their hair turned into deadly upholstery.

7

Magic in the Stone

Tales of rubies and their magic haunt the annals of myth, folklore, and even religion.

The first use of these magical stones was for their vaunted healing and protective powers. For example, Sri Suta Goswami, one of the great commentators on the great Indian epic *Garuda Purana*, states, "Pure, flawless gems have auspicious powers which can protect one from demons, snakes, poisons, diseases, sinful reactions, and other dangers." However, he warns, "flawed stones have the opposite effect." A historical reason for the powers of rubies may partly stem from their extraordinary hardness, which defeated most cutting tools. Surely, only magic could be responsible!

Enough folklore has adhered to the ruby to make its magical powers seem endless. Grind one up and put it on your tongue to cure digestive ailments. Simply wearing a ruby energizes the blood, heals heart and liver dysfunction, cures snakebite, and even prevents drowning. For the Chinese, it meant long life. According to the ancient Indians, if you are in a hurry to boil some water for your tea, simply toss a ruby into a pot and the water in it will boil instantly. In like manner, the Greeks thought rubies would melt wax.

Since Europe produces few gemstones of any sort (and no rubies) it's no surprise that when these exquisite and rare stones arrived during the Middle Ages they were invested with even more magical power than they had acquired in the lands of their birth. Who among Europeans could say where they came from or how they derived their power? Here is a case in point. The beautiful Elizabeth, wife of Franz Joseph of Austria, believed in the power of ruby as a good luck charm. It was for her, certainly. On September 10, 1898, when she forgot to wear her ruby ring, she was assassinated by an Italian anarchist in Geneva, Switzerland. His weapon was a needle file, with which he punctured her heart. Her last words were, "What happened to me?" In her youth (she was 60 at her death) the willowy Elizabeth was renowned as being one of the great royal beauties. She was noted for her masses of hair (taking three hours a day to arrange) and her 20-inch waist, which she maintained by a long-term starvation diet. She even had her picture in an 1884 issue of *Vanity Fair* magazine in her riding habit (she was well known as a skilled equestrienne). Afterward her murder, the assassin, Luigi Lucheni, is reported to have said, "I wanted to kill a royal. I didn't care which one." In point of fact, he had earlier attempted to kill a prince from the House of Orleans.

THE BREASTPLATE OF JUDGMENT AND OTHER BIBLICAL RUBIES

According to Exodus 28:15–30, 12 stones adorned Aaron's gold "breast-plate of judgment." Each stone was marked with the name of one of Jacob's sons: "The stones will correspond to the twelve sons of Israel name by name, each stone bearing the name of one of the twelve tribes engraved as on a seal."

Although each stone is given a name, it is sometimes difficult to equate the ancient Hebrew designations with their modern equivalents. The word sometimes translated as "ruby" is listed as the first stone on the first row, possibly attesting to its importance. The original Hebrew lists the stone as an *odem*, whose meaning is a mystery other than that it was some sort of red stone. And again, any red stone was a ruby for the ancients. *Odem* is rendered as "ruby" in the New International Version and New American Standard Bible, but as "sardius" in the King James and Revised Standard Version. Another suggestion is "red jasper." Still another is "sardonyx." Modern Hebrew, however, uses *odem* to refer to the ruby.

According to some Talmudic legends, the ruby stood for the tribe of Naphtali, the sapphire for Dan, and the emerald for Reuben. Other accounts claim the ruby for the tribe of Reuben and suggest that the ruby was the fourth stone on the breastplate, not the first. But the fourth stone is often supposed to refer to the garnet. The word "carbuncle" is also sometimes brought up. This word simply designates any red, glowing stone, perhaps ruby, perhaps garnet. There is no real way to tell now.

There are good many problems with the entire account of the biblical breastplate. For one thing, the prospect of engraving a ruby is truly daunting—it is much harder than steel and certainly a lot tougher than whatever implements the ancient Israelites might have had. And where would the ruby have come from? Certainly not from the Egyptians, where the Israelites got the rest of their loot. Rubies were not mined in Egypt, and the Egyptians looked upon the color red with something akin to horror. They preferred green. However, it should be said that King Tutankhamen's tomb did contain a few red stones (cornelian and red agate).

Whether the breastplate ever existed or not and what it was made of is a rather moot point. The legends accruing about it are fascinating. For instance, one old story claims that when the anger of the Lord God was upon the people of Israel, the stones in the breastplate darkened, but when he was in a happier disposition, the jewels glowed brightly, thus making the breastplate of Aaron the original mood ring. Supposedly, the sapphire (a heavenly stone) was particularly attuned to the Lord's temper.

The verses in Exodus are not the only biblical mention of rubies, although the others are all metaphorical. The Book of Proverbs proclaims, "A capable, intelligent and virtuous woman, who is he who can find her? She is far more precious than jewels and her value is far above rubies." We understand from this remark that in the ancient world, rubies were the ultimate criterion of value. It should be mentioned, however, that again the translation is not completely certain. Some believe the word is better translated as "coral," "pearl," or even just a bright, reddish stone. Many supposed "rubies" were probably red spinels anyway. It has always been a somewhat of a mystery, at least to me, as to why spinels are held in such low esteem in comparison to rubies. They are almost as hard, just as handsome, and are actually rarer.

In like manner, the ruby may or may not make an appearance in the New Testament, depending upon how liberally some of the words applying to precious stones may be translated. A glimpse of the problem can be noted as far back as Andreas Bishop of Caesurae (431–506 perhaps, or as late as the tenth century, depending on whom you believe. At any rate, he lived a long time ago). Andreas was making a valiant attempt to connect the metaphorical stones of New Jerusalem with what he believed were the literal ones of Aaron's breastplate. Andreas actually used the word carbuncle, and no one really knows what he was referring to. Ruby may be the best guess, although sardonyx also has its supporters. What he actually said, in writing an explication of Revelation 21, verse 19, which listed the foundational stones of the New Jerusalem, was, "The chalcedony [the fifth stone] was not inserted in the high priest's breastplate, but instead the carbuncle, of which no mention is made here. It may well be, however, that the author designated the carbuncle by the name chalcedony." Thus Andreas was assuming the "sardonyx" on Moses' breastplate could be translated "carbuncle," which I and others suggest can be translated as "ruby." Sardonyx simply doesn't have the cachet of "ruby." The word carbuncle is actually related to the word *carbo* or glowing ember. It refers to any red, glowing stone, but commonly a ruby, the most valuable of the lot. It especially refers to a ruby that strongly fluoresces, giving it the appearance of a glowing ember.

Andreas associated St. Andrew with rubies, although his reason for the match up is rather inventive. "Andrew, then, can be likened to the carbuncle," he writes rather desperately, "since was splendidly illumined by the fire of the Spirit." This is a reach, but a good one.

MAGICAL BIRTHSTONES AND OTHER CALENDRIC ASSOCIATIONS

The modern concept of birthstones originated in eighteenth century Poland. The idea originated among the large Jewish community there, and from the beginning there was an obvious connection with the 12 stones on the breastplate of Aaron, 12 tribes of Israel, and the 12 months of the year. It only made sense to link them together somehow. The Jewish historian Josephus was the first on record to draw the connection between the various series of 12 stones.

It was only gradually that the idea came about that certain stones had affinities for those born during certain months. Which stone went with which each month however, was a matter of dispute. The fact that astrological signs intersect with the months rather than coincide with them does not make the determination of which stone is right for which person any easier.

In Poland, the ruby was assigned as a birthstone to those born in the month of July.

Today, indeed, the ruby is usually considered the pre-eminent gem of the summertime. According to an anonymous rhyme:

The glowing ruby shall adorn
Those who in warm July are born;
Then they will be exempt and free
From love's doubt and anxiety.

The Hindus assign the stone to August, although in most of India it is summertime all year round. Indian seasons are more generally divided into "dry" and "wet," and it's anybody's guess as to which is worse. The National Association of Jewelers has assigned ruby to July, and that should make it official. However, it is not universally agreed that rubies belong to July or even August. Some ancient Jews, Romans, Poles, Arabs, Italians, and Isidore of Seville assigned the ruby to December, as do some Russian and modern lists. The switch to July, the hottest and most fiery month of the year, has been recent, but logical. A few lists assign it to April. Furthermore, it is sometimes considered a "talismanic gem" for March.

The modern list was invented at a meeting of the National Association of Jewelers in Kansas City in 1912. It was at this meeting that the jewelers chose the ruby for July rather than the more widely accepted identification of the ruby with December, thus agreeing with the Poles rather than with everybody else. Their interest was no doubt mainly commercial, but the idea has a widespread appeal that there is some mystical, symbolic connection between a birth date and a precious stone. In a broader context, the ruby is considered the gem of summer, as emerald stands for spring, sapphire for autumn, and diamond, of course, for winter.

However, "birthstones" are really borrowed from one or another various astrological systems. The zodiacal sign changes about the 22nd of each month, and one's "birth sign" is traditionally associated with that rather then the month in which happened to be born. Western astrology mostly depends up on sun-signs; however, other systems, sometimes called sidereal systems, are based on the position of the moon. However, both systems agree that each sign of zodiac is ruled by one particular planet. (This is all getting rearranged nowadays, as even astronomers are arguing about the definition of a planet and whether or not Pluto and outlying bodies are awarded the title of planet.)

For example, most of July falls under the astrological sign of Cancer, but the last week or so of that month belongs to Leo. And the last part of June belongs to Cancer, too. Is ruby the stone for July only, or do those born in the last week of June and the first week of August have equal claim? In short, are birthstones calibrated with the months of the year or the astrological signs? It has also been suggested that everyone is supposed to wear the proper stones during the month—or sign—we are passing through, rather than be stuck with the same gem all year round.

In Hindu and Vedic astrology (Jyotish), ruby was the gem for Capricorn (December 22–January 21). Other astrologers argue that the stone is "too powerful" for Capricorn, Virgo, or Pisces, who would be better off wearing a substitute stone. In other systems it is Aries and Libra who should stay away from it. In some Arabic lore and in Babylonia, it was declared that the ruby was under the influence of Taurus, but this is definitely a minority opinion. One Spanish writer, Gaspar de Morales, connects the ruby with the constellation Hyades and the particular star Aldebaran.

Thus, the ruby was like the sun, a light unto itself. Sometimes it was even called the "lamp-stone." This belief was not limited to India. An Emperor of China, it was said, used ruby light to illuminate his chamber. (The King of Ceylon must have owned a bigger one, because his was said to light up the entire palace.) Red has always been a lucky color in China, and its red lanterns are famous. In ancient China, officials wore badges made from gems to denote their rank. Mandarins of the highest rank wore, of course, rubies, red tourmaline or another red stone. (As in ancient Egypt, the color was more important than the actual species of stone.) In Greek legend, a female stork repaid the widow Heraclea for a kindness done. We know the story from the works of the rhetorician Aelian (fl. c. 220), the "honey-tongued." Heraclea had taken care of a baby stork that had fallen from its nest and broken its leg. One night in the following year, returning from its annual migration, the stork dropped a brilliant ruby onto Heraclea's lap, while the old lady sat snoozing in the door. The stone was so bright that it illuminated the entire apartment and woke her up. Whether that ruby became prized for the light it brought or its cash value is unclear, but at any rate, it was a precious gift. This may have been the white stork, which migrates annually from Europe to Africa, especially Kenya; if so, it was undoubtedly a Kenyan ruby, probably the first "on record." Aelian has a lot of such stories, gathered into his *De natura animalium* (*On the Characteristics of Animals*). The point of most of his fables is to contrast the behavior of animals with that of humans, sometimes to the distinct benefit of the animals.

A similar tale about the light-producing qualities of a ruby comes from ancient Jewish legend. Abraham kept his wives shut up in an iron city; however, he generously gave them a bowl of rubies to provide light.

In the Indian system, there are nine gems, the *navaratna*, each of which relates to a heavenly body. All are considered to be equally sacred. However, at the center in the position of honor is the ruby, which is the gem-embodiment of the sun, and which has absorbed its powers, as the epic *Garuda Purna* relates. (The Burmese entirely agree with the Indians in handing the ruby the place of honor, the first among gems.) The ruby is considered to be associated with solar energy, and with its association of fire, fatherhood, masculinity in general, royalty, and political power. It is indeed the stone of kings. In the same tradition, all other gems—the diamond to the east (Venus); the pearl to the southeast (Moon); sapphire to the west (Saturn), yellow sapphire or topaz to the northeast (Jupiter); emerald to the northwest (Mercury); Zircon or jacinth to the southwest (Rahu, the ascending Moon); cat's-eye to the north (Ketu, the descending Moon); and coral to the south (Mars) must circle around it and pay it homage.

The planetary-gem connections are somewhat different from those in the West, where diamonds are typically associated with the Sun and emeralds with Venus.

Today, the ruby is also the traditional wedding gift for the 15th and 45th anniversary, another idea invented by jewelers to make people spend money. In more fiscally conservative times, the ruby was reserved for the 45th anniversary, but as these are less common than those celebrating the 15 year mark, jewelers decided it was the better part of business acumen to suggest that the ruby be given earlier as well.

Precious stones have their weekdays as well as their months. The day most often associated with the ruby in India is Sunday, as the stone is associated there with the sun, on account of its burning heat. If it is associated with warlike Mars, the red planet, on account of its bloody appearance, then Tuesday is its lucky day. One astronomical myth about rubies is that they are sparks struck from the red planet, and their fire will never be quenched until the world itself grows cold.

Many crystal workers claim that rubies' strongest magic is worked on Friday. This is because Friday is traditionally associated with Venus, and Venus is the goddess of passion, and ruby is the stone of passion. Star Rubies should be worn on Wednesday, although the reason is unclear. (No matter what the day, its most lucky hour is said to be between 5 and 6 o'clock in the afternoon.) Indian folklore insists that Sunday, Monday, or Thursday are the ideal days to buy the stone, especially if you do so when the moon is in an ascending cycle.

According to folklore, also the ruby is a lucky stone for women named Rose or men named Roland. However, a bit of sexism creeps in when it is averred by some of the ancients that rubies (or red clothing) worn by a man signifies nobility and courage. However, red on a woman simply symbolizes her pride and obstinacy. For men, red reveals itself as the color of warlike valor; for women it simply makes them vengeful.

PROPHESYING, GOOD LUCK, HEALING, AND PROTECTIVE RUBIES

Symbolically the ruby represents liberty, generosity, passion, and divine power, and its attendant powers reflect these values.

It is also a great healer. In Indian lore, the royal ruby, the stone of the sun, is a preeminent healing stone. (If you don't have a ruby, a red spinel or a red garnet will do. But the unspoken consensus is that a ruby is really best.) The theory is that the sun acts like the thalamus in the body. Just as the sun is the controller of the solar system, the thalamus is in control of many physiological processes in the body. The ruby empowers the thalamus and thus aids in the general health of the body. (According to some schemes, ingesting powdered ruby keeps the body from corruption after death.) For the living, powdered ruby was said to banish plague and pestilence (possibly typhus), and cure the wearer of vain, foolish fancies. Taking 10 to 40 grains of powdered ruby was also said to "sweeten the sharpness of the humors" and strengthen the vital organs.

In Indian lore, also, the ruby is sometimes associated with opening the Sacral Chakra (the second chakra), but it is equally associated with the fourth or Heart Chakra. It heals the heart, both physically and emotionally.

Among some contemporary crystal healers, ruby is said be a carrier of the "red ray." It symbolizes masculine virility and the Sun. In some schemes its planet is Mars, the "red planet," although in classic Ayurvedic medicine coral goes with Mars. Ayurvedic practitioners used ruby to heal ailments of heart, spleen, skin, hypertension, brain, bones, and eyesight. In older times, for the ruby to be effective, it had to be "bruised in water." And if you went about your property, touching the four corners of the house, courtyard, or vineyard with a ruby, the said areas would be preserved from lightning or worms.

In contemporary times, rubies are said to be helpful for Alzheimer's sufferers, and are supposedly indicated for infections, fevers, bile, acne, low energy, poor circulation, edema, and arthritis. Rubbing a ruby on the skin is said to maintain its youthful, dewy appearance. However, a dissenting view is occasionally voiced that rubies actually disturb the circulation of the blood and arouse anger in the wearer, although Ayervedic practitioners use it to cure hot temper and impatience. The action the ruby is reputed to have depends upon whether one subscribes to the like cures like philosophy, or to the opposite belief that "hot-blooded stones" simply increase the same tendency in their owners. However, since both theories produce equally unimpressive results when actually tested, it may not matter.

Grinding a ruby to powder and placing it on your tongue will cure indigestion, although I daresay most people will put up with a little tummy ache before grinding up the family heirlooms. The original source for this bad idea can be traced to the thirteenth century Kashmir physician Naharari, who used ruby as a remedy for curing flatulence. It is well known that practically all the ancients believed in the healing power of red stones. All (ruby, spinel, garnet) were said to cure bleeding and inflammation, for obvious reasons of sympathetic magic.

The alleged healing properties of the ruby have never been limited to physical disease. Its healing properties include nurturing, spiritual wisdom, prosperity, and protection from distress. Some traditions say that rubies protect against evil thoughts—including one's own. They conquer fear and worry. In the realm of mental health, ruby helps its wearer overcome temptation and promotes psychological stability. If worn on the left hand, contend some contemporary "crystal healers," it brings good luck as well.

The love the ruby is said to induce is a combination of sensual love and agape, with a decided balance in favor of the former. If a diamond engagement ring symbolizes eternal love, a ruby ring signals fiery passion. However, while rubies represent love and passion, there is a catch. Tradition tells us that any attempt to use a ruby to force love will backfire upon the user.

Ruby also controls the sexual appetite, which may seem like an odd association for such a passionate stone. However, this seems to be a modern view. More anciently, the Italian Hierononymus Cardanus or Gerolamo Cardano (1501–1576) merely opined that ruby would cure "vain thoughts." Cardanus was admittedly no expert on gems, although he thought he was. He was more famous as a mathematician best known for his formulas to solve cubic equations (Cardano's Formula). He also wrote a good deal of philosophy and natural science, but made it clear that his work was for the upper classes only. This was all right, as a good deal of it was totally wrong.

The ruby is also supposed to confer protection upon its rightful owners by warning of danger. (It grows darker when danger is nigh and returns to its original color once that danger is past.) One story, recounted by C.J.S. Thompson in *The Mysteries and Secrets of Magic*, concerns one Gabelschoverus:

On the fifth day of December 1600 I was going with my beloved wife Catherina from Stuttgart to Caluna. I observed on my way that a very fine ruby which I wore mounted in a gold ring which she has given me, lost repeatedly and each time almost repeatedly its splendid color, and that it assumed a somber blackish hue which lasted several days; so much so, that being greatly astonished, I drew it from my finger and put it in a casket. I also warned my wife that some evil followed her or me. And truly I was not deceived, for within a few days she was taken mortally sick. After her death the ruby resumed its pristine color and brilliancy.

A similar story, however, ended up with a scientific explanation. According to a tale related by Johann Jacob Spener about a jeweler and his ruby ring:

One day, after having washed his hands, this man sat at a table, when, glancing at a ruby ring he wore on his finger, he remarked that the stone, which usually delighted his eye with its splendor, had lost its brilliancy and become dull. Since he believed what others had related to him, he was firmly persuaded that some misfortune threatened him.…A fortnight later one of his sons died.

The jeweler checked the ruby ring, which he had put away, and found that the stone had regained its brilliancy. The jeweler resumed wearing it, only to find shortly after washing his hands that the stone again turned dull. This time however, he undertook to investigate the matter more carefully, and discovered that the dullness was due to a bit of water getting between the backing and the stone and dulling its effect.

Since royalty were generally the only people who could afford a ruby, they got more protection than most. The stones were supposed to absorb the evil rays from malevolent stars and, by concentrating them, turn them into good rays. Cracks in the stones were not only unpleasant to look at, but actually dangerous since they let the bad rays through.

The protective powers of rubies were especially noted by Sir John Mandeville, who, in a fourteenth century treatise attributed to him (*Le Gran Lapidaire de Jean de Mandeville*), praised the ruby for its miraculous protective abilities. He wrote that the owner of a good ruby would be able to live in perfect peace among all men, that neither his land nor riches would be removed from him. He added that the stone would also guard his house, vineyards, and fruit-trees from tempests.

Mandeville is best known for his romantic *Travels*, a topic on which he was expert, having left England in 1322 on pilgrimage to Jerusalem. According to his own account, his journey took him to Egypt, where he fought as a mercenary in the sultan's wars against the Bedouins. Then he visited the Holy Land, then India, then the interior of Asia and China, where he claimed he served for 15 months in the army of the Great Khan of Mongolia. After an absence of 34 years he returned in 1356 to France to enjoy a well-deserved rest.

It used to be thought that there never actually was any Sir John Mandeville, and that his works were written by somebody else named John or Jean. However, scholars are now coming around to the idea that Sir John Mandeville existed after all, just as he claimed he did. Whether or not he ever left France is another story. In addition, the authorship of the gem treatise is not even so certain as that of the travelogue.

Mandeville's writings have an incomparable charm, although their accuracy is slightly suspect. Let's take what he said about Egypt, which he described in this way: "Egypt is a long country, but it is straight," by which he meant "straight" as in "straight and narrow." This is quite true, as only the part of the country near the Nile is of any use whatever. More amazingly, according to John, Egypt is the home of the Phoenix. "There is none but one in all the world," he wrote earnestly, and fairly accurately, being off only by one. "And he cometh to burn himself upon the altar of that temple at the end of five hundred years; for so long he liveth."

The ruby is claimed to bring confidence, courage, and self esteem. The Hanza people of the Kashmir frontier, for example, were said to shoot ruby bullets during their 1892 rebellion against the British and Kashmiri imperial troops. (Other sources pooh-pooh this notion, claiming that the bullets were made of mere garnets.) In either case, the idea apparently was that a blood-colored "bullet" could inflict greater damage.

A related story claims that ancient Burmese warriors used rubies as "bullets" for their blowguns (they are certainly harder than lead or silver). Burmese soldiers also inserted rubies in their skin, an operation which they believed made them invulnerable and perhaps even immortal. While there's no proof of this, it does seem that ruby lived up to its reputation of instilling self-confidence, for the courage and indeed recklessness of these soldiers was widely remarked upon. It remains uncertain, of course, as to whether the courage of the ruby-bearing soldiers was a direct gift of the stone or of their belief in its bullet-proofing powers. Perhaps it all amounts to the same thing. For those soldiers unwilling to insert rubies into the more tender parts of their flesh, it has been suggested that an earlobe will also work nicely.

It is also said that most unicorns bear a mystical ruby at the base of their horns. This ruby is the source of the creature's tremendous power. This myth is probably of Levantine origin, brought back by the Crusaders. According to a medieval poem by Pfaffen Lamprecht, Queen Candace sent such a unicorn to the Conqueror, and the value of the gift was more in the ruby than in the magical animal itself! Wolfram von Eschenbach's *Parzifal* reiterates the same legend:

We took from underneath his horn
The splendid male carbuncle-stone (*karfunkelstein*)
Sparkling against the white skull-bone...

In tenth century China, dragons and snakes were carved into the surfaces of rubies, supposedly in order to increase the flow of money and power to their owners. A like story is mentioned in "Ragiel's" pseudonymous thirteenth century *Book of Wings*, in which it is stated that if the "beautiful and terrible figure of a dragon" is carved upon a ruby, it will increase the owner's goods of this world, and as an add-on, make the owner happy and healthy.

There is also a rumor that Madame de Pompadour kept on her person a large ruby carved in the shape of a pig, which she wore for good luck. If so, she didn't mention it in her will. If it did exist, it probably didn't bring her much luck anyway, as she died at age 42.

Enter the Seeress of Prevorst, a woman whose real name was Frederike Wanner Hauffe, born in 1801; she lived a mere 29 years. Frau Hauffe was a well-known "sensitive" of her day. She was famous from childhood on for seeing ghosts and demons, often in rather unlikely places, and one of whom gave her legal advice. When she was 19 years old, her family fixed up a match for her with Herr Hauffe, and the idea pleased her so much that she spent entire days weeping ceaselessly. It did her no good however, and the wedding went forward as planned. Not surprisingly, perhaps, her health continued to decline, despite the vast amount of bleeding, magnetic therapy, mustard poultices, magical amulets, and even a special machine designed by herself, administered on her behalf. She began to hemorrhage with alarming frequency, her teeth fell out, and she developed a deep dread of all men.

One of the more interesting aspects of her case was the intense response she had to various minerals. If she were handed a bit of granite, flint, or porphyry, she remained unaffected. However, fluorspar gave her diarrhea. She found rock crystal and other clear stones invigorating (sometime to the point of causing convulsions), but sapphire put her to sleep. Sapphire's sister, ruby, however, had a completely different effect. Merely holding a ruby made her tongue grow cold and heavy, so much so that she could only utter the most incoherent of sounds. Gradually, her fingers and toes also took on a chill, and she was seized by a fit of the most violent shivering. Soon afterward, though, she was restored to a sense of the greatest well-being, although there always lurked the dark thought that the symptoms might reverse themselves once more at any time. This did indeed occur, eventually, and death was only a relief to her.

The boundary between the use of gems as medicine and their use as talismans is a porous one. One may have derived from the other, and which came first is problematic. In addition, ancient cultures did not readily distinguish between natural diseases and conditions brought on by the evil eye or accident. The use of precious stones to heal disease and avert diseases shows up in almost every culture studied. In some cases the stones are worn, in others ground up and swallowed, but the purpose was the same. In many cases, the use of the stone was dictated by its color rather than its chemical composition, which, after all, the ancients had no way of knowing.

Red stones suggested blood, and thus they were believed to stanch bleeding and heal the heart (both physically and metaphorically). Sir Jerome

Horsey, a sixteenth century English emissary to Russia, recorded that Ivan the Terrible (more politely known as Ivan IV) was a great believer in the medical power of gems. Pointing to a ruby, says Horsey, the mortally ill czar cried out, "Oh! This is most comfortable to the heart, brain, vigor, and memory of man, clarifies congealed and corrupt blood." Looking sadly at this great collection of gems, he said, "All these are God's wonderful gifts, which he secretes in nature, and yet reveals them to man's use and contemplation, as friends to grace and virtue and enemies to vice. I faint, carry me away till another time." Alas, it was too late for the stones to be of any use to the terrible czar, if indeed they ever were.

Recently there has been something of an attempt to rehabilitate Ivan's reputation, with some historians suggesting that perhaps he wasn't all bad. To his credit, he compiled Russia's first code of laws in 1550, established Russia's first printing house, and reformed the government. He also annexed Siberia, although it is not recorded that he asked for the consent of the Siberians before doing so. None of these accomplishments, however, can quite wipe away his odd behavior. He used to go around with an iron-pointed staff, which he used to attack anyone who offended him. This included his own son, whom he clubbed to death in 1581. He was extraordinarily fond of torture too, and took great delight in designing agonizing deaths for people sentenced to be executed. He said he wanted to reconstruct the sufferings of hell for the doomed and apparently didn't want to wait until they got there. He also instituted Russia's first secret police, whose main job it was to rape people and help with the torture. Defenders of Ivan say it wasn't really his fault and that his obvious insanity was a result of syphilis.

When Ivan's tomb was opened during some renovations undertaken during the 1960s and his remains were examined, it was discovered that the body had very high levels of mercury, indicating that he had probably been poisoned, although some suggest the mercury was meant as a treatment for syphilis. In any event, his own death was a rather terrible one in its own right, although nothing compared to the doom he arranged for others. He actually got off rather easily.

In the early part of the twentieth century, William T. Fernie, M.D., a believer in the almost miraculous power of gems, ventured that a ruby "by virtue of its metallic oxides, of iron, copper, and chromium" may "renovate the bloodless patient, bringing back the rich hue of convalescence." He had equally odd things to say about other gemstones. More modestly, contemporary crystal healer Brenda Knight suggests that wearing a ruby on the middle finger will awaken one's "inner and outer beauty."

Indian Ayurvedic medicine also makes use of ruby cures. Ruby *bhasma* is supposed to be a sweet-tasting aphrodisiac, heart tonic, and digestive and metabolic stimulant. It is also used to treat tuberculosis. Because the ruby is considered a hot, dry stone, it is effective in treating illnesses that come about through an excess of cold and dampness, such as flu, colds, anemia, and so on.

However to be really effective, the ruby must be eye-clean, 3 to 5 carats, and set in gold (or gold and copper, with a minority report suggesting silver). It should be worn on the ring or little finger, but whether the best results are obtained on the right or left hand is disputed. In any case, the new owners should first put on the ruby on Sunday, Monday, or Thursday, at sunrise during the "bright" half of the month when the moon is waxing. Ideally, according to the principles of Indian astrology, it should be put on when the sun is in Leo in the constellation of Pushya, although it could also be put on during Sagittarius. The following mantra should be recited:

Aum grinih suryaya namah!

It's actually best to chant this mantra 108 times a week, taking special care to recite it at dawn, noon, and sunset on Sunday.

Before actually wearing the stone, however, one should bathe it in pure, unboiled cow's milk or purified water (or water from the Ganges River, according to Indian belief) and waft some consecrated incense around it. There is a disagreement about how long or much a ruby needs a milk bath. For some authorities, one dip is sufficient. Others recommend 100 times to be safe. It can then be venerated with flowers and incense. While diamonds are generally considered to look exceptionally good when matched with rubies, practitioners of Indian astrology warn against this practice, believing that rubies (or their substitutes) should not be worn with a diamond, blue sapphire, cat's eye, hessonite, or their substitutes, because Venus, Saturn, Rahu, and Ketu, planets that these stones represent, are not compatible with the sun. Using these gems in a Navaratna is an exception to this rule.

Rubies have their role in religion as well as in magic. Perhaps no land on earth accords them as much spiritual value as India, where rubies are regarded as the chief among all jewels.

Rubies are sometimes also said to protect wealth, and as long as one retains a bit of ruby, wealth will not be lost. On the other hand, hanging onto a ruby is tougher than it may sound, because another magical (and disconcerting) aspect of rubies is the way they can disappear. In fact, the location of many of the world's greatest rubies—the Dragon Lord, the Chhatraputi Manik, the Nga Mauk, the Maung Lin, the Tagougnandaing, the Pingu Taug, and the Mandalay rubies—are presently unknown. Most are probably in the hands of private collectors—somewhere. Some many have broken up, and a few genuinely misplaced. A few were bought up by the incredibly rich and royal, as we have seen.

Rubies are able to induce strong guiding dreams. And dreaming about rubies themselves is fraught with magic. It may signal the arrival of unexpected guests, although I suppose if you know this, the guests will no longer be completely unexpected. If you want to avoid bad dreams, whether of unexpected guests or not, simply slide a ruby under your pillow. Modern crystal workers say that dreaming of a ruby brings joy, good fortune, and success in business and love.

On the supernatural level, rubies chase away evil spirits of the dead and even the demons from hell, should they wander too far from their fiery pit. As an amulet, rubies are said to overpower sorrow and grief. As a ring, a ruby imparts wisdom, health, and prosperity. Wear one as a brooch on your left side and you will live in peace. Embed one in your own flesh, as the ancient Burmese warriors did, and you will be protected from danger in battle.

But beware. When rubies change color, disaster will strike. It is said that Queen Catherine of Aragon wore a ruby that turned dark and dull one day before Henry the Eighth announced he was divorcing her. (His passion had likewise faded.)

In 1583, the Swiss alchemist Leonhard Thurneysser (1531–1595) believed that "ruby brings joy, and strengthens the heart." Thurneysser was more than an alchemist, however. He was a goldsmith, miner, printer, astrologist, and uroscopist. In the last capacity he hypothesized that urine distillates and their residues should be burnt in order to define their composition from the color of the flame. Burning urine wasn't as bad an idea as it may first appear (a modern flame photometer measures the concentration of sodium and potassium in material such as urine by burning it), but over the years, Thurneysser's theories became more and more abstruse, although he did get a nice note in the September-October edition of the 2003 *Journal of Nephrology*. Rubies were not mentioned, however.

Thurneysser worked hard at his alchemical studies, but was eventually forced to flee Basel when he was caught selling gold-covered lead as pure gold. He joined the army of Albert, Margrave of Brandenburg, was captured in 1553 by the Saxon army, and was put to work in the mines (not ruby mines, however). When released he traveled to England, France, Bohemia, Hungary, Italy, Spain, and North Africa to learn the metallurgical and medical practice of the day. Later, he set up an extraordinary laboratory employing up to 300 people at the Greyfriars monastery in Berlin, which produced saltpeter, mineral acids, alums, colored glass, drugs, essences, and amulets. He also established a printing house that put out calendars and predictions, as well as alchemical and medicinal tracts. As the books frequently included words from foreign languages, he was accused of harboring a devil in his inkpot who dictated everything to him. In spite of, or perhaps because of, the devil, Thurnysser became rich, although he lost most of the money in a divorce from his second wife.

TALES OF HINDU AND BUDDHIST MAGICAL RUBIES

Earlier I spoke of the *Garuda Purana*. Another of these epic Hindu tales, the *Bhagavata Mahapurana*, tells the fate of the magical gem called the Syamantaka Ruby. The story tells us that 5,000 years ago, at the end of the "Copper Age" of Dvapara-Yuga, the god Vishnu made his eighth incarnation as the Lord Krishna. (We are now in the Iron Age or Kali-Yuga, and have steadily deteriorated from the original age, which naturally was the "Golden" one.) As usual, a little trouble developed, something that always tends to occur when the gods get too involved in human affairs. This time, more than one god was involved, also a bad sign.

Enter King Satrajit, a devotee of the ancient Vedic god Surya, lord of the sun. In honor of his years of devotion, Surya passed along a divinely powerful ruby to the pious king. Its name was Syamantaka, and it possessed the magical ability to produce 170 pounds of gold every day for its owner. This ruby was so bright that when the King wore it in a locket around his neck, people thought it was the Sun-God himself.

The trouble really started when the king took it into his head to visit Krishna on his island kingdom of Davaraka. Apparently, this was quite a place ("matchless," according to the text). It was square, measuring a hundred yojanas. A yojana is about eight miles, so we are talking about a pretty large area here. The place was decked in pearls, rubies, diamonds, "and other gems," adds the author dismissively. The whole place blazed like the "meridian sun" in summer. Even here, however, in a city practically drowning with gems, this particular ruby was noticed. ("Drowning" is not really a metaphorical term here. Currently this lost city is the subject of underwater archaeology, and quite a bit of it has been recovered.)

Apparently wanting everyone to be impressed with the gem, he placed it in a temple to be worshipped by some priests he hired just for that purpose. It was not worshipped for its beauty or gold-producing power alone. It was believed that wherever this gem was worshipped there would be no possibility of famine, disease, or pestilence.

Krishna slyly asked the king for the gem, but Satrajit refused. It is always a serious error not to hand over a ruby to a god when asked. Indian lore assures us that offering a ruby to Krishna will ensure the supplicant's good luck; one can only imagine what happens when one actually refuses a direct request. Krishna then suggested that perhaps it might be nice to offer the gem to King Urgasena, the ruling monarch. Again Satrajit refused. All this caused quite a stir among the citizens of Dvarka, who assured Krshna that if only they had had a good ruby, they would be happy to donate it.

However, Satrajit did lend the gem to his younger brother Prasena, who then went hunting. Rubies have many powers, but success in the hunt is not one of them, and in fact, Prasena was forthwith killed along with his horse by a lion, possibly for the stone. The lion was killed in his turn by Jambavan, the King of Bears (or gorillas, by another account), who thought the ruby would make a nice trinket for his little son.

When Prasena didn't come back, Satrajit became extremely distraught, whether because of the loss of his brother or his ruby is not quite clear. Human nature being what it is, people started slandering Krishna and blaming him for the loss of the gem. Krishna proved his innocence by leading a search party, finding the bodies of Prasena and the lion and tracking down Jambavan. He then challenged Jambavan to a fight, which went on for a month before Jambavan finally recognized Krishna and gave up. He then offered Krishna the gift of his daughter Jambavati, and the ruby. Krishna accepted and showed his greatness by returning the ruby to king Jambavan, who in his embarrassment had tried to make a present of it. It was also quite nice of Krishna to marry the daughter of a gorilla. He also married the daughter of Satrajit, Satyabhama. To this very day, an offering of a ruby to Krishna guarantees the sacrificer eternal life.

Buddhist legend too has tales of priceless rubies. Chief among these tales is the Vidhûra Jataka, one of the many Jataka tales (Number 545 in the Pali canon) that tell the story of the Buddha in one of his previous incarnations. In this case, the future Buddha was born as the noble Vidhûra, and the story details how he overcame the evil Rakshasa Punnaka. In the story, Vidhûra was presented with a glorious ruby by the King of the Nagas (snakes) as a reward for deciding a case. The king had been accustomed to wearing the gem around his neck, and his queen Vimalâ noticed it was missing. When the king explained that it was a gift, the queen immediately wanted to know if Vidhûra was the same one she had heard of, who spoke about the Dharma or spiritual law. When the king answered in the affirmative, the queen fell into deep thought. More than anything she wished to learn the Dharma but suspected the Snake King wouldn't allow it. So she devised a stratagem.

She lay on her bed and pretended to be very, very sick. So sick, in fact, that she told the Snake King that she could revive only if given the heart of Vidhûra, lawfully and not by force or guile. "Now who is going to manage that?" demanded the king. "I have no idea," responded the queen, "but it had better happen." With that she turned her face to the wall. Desperate, the King sent out his daughter Irandhati "to look for a husband" who might be able to reel in Vidhûra to the underwater Nagaland. He obviously wasn't going to try it himself.

Irandhati, who was as cunning as her mother, swam up to the world, and strewed a mountain with red flowers to make it look like a "heap of rubies." One doesn't usually catch the best types in this way, and sure enough, Irandhati was able to haul in the evil Punnaka. The bride-price was, of course, the heart of Vidhûra.

Punnaka trotted off to Vidhûra's uncle's place. Or rather, he flew on his magical steed, which was three leagues long and could drive away all enemies. He was temporarily stymied, as he knew that he was unable to conquer Vidhûra (who kept many attendants) alone. He knew he needed the help of a magical, wish-fulfilling ruby, specifically one in the hills of Vepulla, near Râjagriha. (Even today the inhabitants of Burma call rubies Ma Naw Ma Ya, or "wish-fullfilling gems.") The ruby was called Manohara (the Captivating). One wonders at Punnaka's seeming audacity, as the ruby was guarded by 100,000 Kumbhandas. Apparently, they were pretty useless, however. For, upon ascending to the top of Mount Vepulla, he terrified the guardians of the stone with a single look, walked off with the gem, and headed to the land of King Korabya, where Vidhûra served as minister.

Punnaka decided the best use for the ruby was a stake in a game of dice, along with his magical horse, in order to play for the prize of the future Buddha. (To this day, the ruby is considered a good luck charm for gamblers.) However, Punnaka had to talk King Korabya into the value of both the stone and horse by displaying their many virtues. In addition to its other virtues, the ruby seemed to function as a crystal ball, for by merely looking into it, one could see all the cities of the earth and even glimpse the kingdom of the gods. The king finally agreed, saying that if Punnake won he could take everything he had except his life, his queen, his throne, and his white umbrella.

Korabya had a trick of his own up his sleeve—namely, a fairy who in a former life had been his mother and who could be whistled up to help him. The fairy kept interfering in the dice game until Punnaka spied her and gave her one of his famous glares, which chased her away. Punnaka then won the game.

When he announced that as a prize he wished to claim Vidhûra, Korabya was horrified. "Look here," he said, "at the beginning of the game I told you that if I lost I would give you whatever your desired, except for myself, my queen, my throne, or my white umbrella. Vidhûra is like my very self." This seems a good bit like quibbling and in any case is a reminder that one should state clearly what one is gambling for before the game starts, but it was too late for that.

"Very well," suggested Punnaka smoothly, "let's ask the wise Vidhûra himself about this matter. Let's see what he thinks and abide by his decision." Korabya had to consent, as otherwise it would look as if he were going to renege on a debt of honor. Naturally, the future Buddha agreed to go along with Punnaka. Pananka decided that the easiest way to manage things at this point was just to kill Vidhûra and simply take his heart along to the Queen. Because of the coy way the Queen had phrased her request, he didn't actually want to lay hands upon him, but merely frighten him to death. Of course, that didn't work. Future Buddhas are not killed off simply because someone gives them an ugly look.

Vidhûra then spent some time discoursing on the Dharma, and his words so impressed Punnaka that he decided to give up his evil ways. He offered to return Vidhûra to his home, but the Buddha-to-be earnestly requested to be conducted to the home of the Naga King. This was done. Vidhûra preached some more, and then met the Queen, preached again, and offered her his heart and his flesh too, if she needed them, in a rather reverse Shylockian tactic. Of course, this is an example of the infinite compassion of the Buddha, who in most of the Jataka tales sacrifices his life for the greater good. But the Queen confessed she had only wanted to hear the Dharma from his lips and that her desire was now fulfilled.

The happy Punnaka then said, "O wise nobleman, I will repay you for the good you have done me by giving you this Manohara ruby, and this very day restore you to your home." You might think that the future Buddha would refuse the gift, but a wish-fulfilling ruby is nothing to turn down. We have already seen that Vidhûra had rather a weakness for jewels. "Since you are a true friend," he said, "you may give me the ruby and restore me to my home." So it was done. All the major characters became part of the Buddha's last incarnation—including the horse, who transmogrified into the Candaka, the noble white steed of the Buddha, before he attained Enlightenment. (Interestingly, it became part of established Buddhist lore that the ruby represents a human soul about to meet the Buddha.)

To this day in India one of the great spiritual offerings is the *panchratna*, an offering of five precious items: gold, diamond, sapphire, ruby, and pearl.

For a while (usually 10 or 12 years) these precious gems maintain a residence in the temple; after that, it is believed that their spiritual value has largely been exhausted and that they can be sold for the upkeep of the temple. But all this is not as easy as it sounds. According to an article in the June 30, 2006, edition of *The Hindu,* an online edition of India's national newspaper, the Secretary of the Podhu Dikshithars of Sri Sabhanayakar Temple (Natarajar Temple), Chidambaram, announced that any devotee desiring to make jewel offerings to the temple must follow a prescribed procedure. The temple would not accept any offering unless the devotee made an application asking to do so. At that point, the general body of the Podhu Dikshithars would consider applications and make a decision. If the application was accepted, the Dikshitars would specify the size and weight of the ornament. The ornament would be accepted only if it conformed to the specifications prescribed. It's important to keep the god satisfied and it was quite clear they weren't going to take just anything. Apparently, there was also a bit of scandal, with crooks and con artists going about gathering gold and jewels form naïve Hindus on the pretext of offering it to a god, and then making off with the riches themselves.

Burmese oral tradition also makes much of rubies. In one ancient story, borrowed from Hindu mythology, Prince Sudhan, married one of the beautiful Keinmari sisters, Manawhari (or Manohara). The prince himself was rumored to be as handsome as the autumn moon, so it is certain they were a fine looking couple. This part of the story is not as romantic as it may seem, however, since the girl was captured with the help of a dragon and presented to the Prince as a gift. At any rate, he fell instantly in love, but a jealous courtier managed to lure the prince away with news of a far-off rebellion, and then told the king that the princess had to be sacrificed. Manawhari escaped, however, and left her ruby ring as a token with a hermit, which was then picked up by the prince who was searching for his beloved. The two were then reunited.

CRYSTAL POWER TODAY

If you believe that curative powers of ruby (and other precious stones) are a relic of a bygone era, reconsider. "Crystal healing" is on the rise, and rubies are part of the picture.

Modern psychic crystal healers use rubies to protect against both headaches and "psychic attacks." To maximize the ruby's powers, it is necessary to set the stone in a ring, brooch, or bracelet and wear it upon the left side.

It is stressed among those practicing crystal healing, however, that an inexpensive red stone like garnet may be substituted for ruby. Indeed, according to many, an inclusion free lower-priced garnet is more efficacious than an included ruby. In like manner, a deep pure-red garnet is preferred to a pale red ruby. A certain Jolm Schroder, writing in 1669, had his own standards, remarking that "you may try the goodness of the Rubine by the mouth, and the tongue, for the coldest, and hardest are the best."

Naturally, the best use of a ruby is to heal a broken heart, but practitioners of crystal medicine claim that the gem can be used for any heart disease. Because of its "radiating warmth," the stone supposedly energizes the entire chest area, and even speeds up the metabolism, presumably thus working as a diet aid. It is even supposed to stop internal bleeding, help out the spleen, and act as an aphrodisiac. This is, of course, opposed to the belief that the ruby was supposed to cure sexual desire.

The idea that rubies and other gems have attendant angels is gaining some currency among modern crystal healers, as angels are very trendy. For D.J. Conway (*Crystal Enchantments*), ruby's attendant angel is Uriel or Auriel. For Doreen Virtue (*Crystal Therapy*), the angels are Jophiel and Michael.

While some crystal healers maintain that rubies can do all these things by merely being held in the hand or laid against the affected body part, there are other methods of administration. One is the so-called "ruby elixir." The origins of the ruby elixir go back to old India, where it was claimed that the elixir could cure a variety of ills. Apparently, however, this ancient elixir received its name merely because it was red, and not because it was made of liquefied rubies.

The elixir manufacturers make great claims for their version of an ancient recipe. They say that because the ruby is a "sun stone," they initialize the key phase of making the elixir on the summer solstice. They say this elixir works with rubies from India, grinding the gemstones into "nanometer-sized" clusters and using clear quartz "nanoclusters" as amplifiers. The clusters are then dissolved in a vegetable-derived solvent imbued with the "spiritus mundi." How this is done is not explained, or even what exactly is meant by "spiritus mundi," or "the spirit of the world." The remedy is designed to work with either your "energy body" or your "astral" body. The effect, manufacturers promise, will be immediate. Emotionally, it will cure you of egotism, or even help you recover from rape, incest, or any other sexual trauma. Your brain wave patterns are supposed to change while you are wearing the stone, thus aiding your recovery.

Some crystal healers warn, however, that people with pacemakers, or people who use conventional medicine, should consult with their doctors before embarking on a course of "ruby therapy." Indeed.

At any rate, the elixirs are extracted by circulation for six weeks in an incubator. The manufacturers explain that this is precisely the method used by the Cathars and the "Dutch School" of Western Alchemy. But nowadays we can improve on medieval methods. The modern elixir is pH adjusted and treated with cold plasma technology for further empowerment. This, then, is the "mother tincture." One part of the mother tincture is diluted with nine parts of alcohol and water. At that point the tincture is treated as a homeopathic remedy. This means that the remedy is repeatedly diluted and shaken ("succussed"). The more it is shaken, the more it is "potentiated" by some mysterious transfer of energy. The remedy itself is chosen to match the symptoms of the patient (based on the "like cures like" concept, although the Latin phrase, *Simila similibus curantur* sounds more impressive. The ruby elixir is shaken to a potency of 1X, a rather low potency in homeopathic terms, as it diluted only to a tenth its original strength (recall that in homeopathy the more it's diluted the stronger it is). The whole homeopathic system was developed by the German physician Samuel Hahnemann (1755–1843) and has an enormous number of adherents, although there is no scientific proof that it has anything other than a powerful placebo effect. The final formula is 25 percent alcohol by volume, also. Since it is not highly diluted, the remedy carries more ruby "nanoclusters" than it can hold in solution. The customer is directed to give the medicine a final shake before swallowing. The suggested dose is three or four drops in water each day between meals.

8

Synthetic Rubies, Fake Rubies, and No Rubies at All

The ancient Indians had their own way of distinguishing the real ruby from fake stones. Simply place the subject stone upon an unopened lotus bloom; if it's the genuine article, the lotus will open and bloom in a short time. The question of how short a short time should be is not addressed. Most lotus buds will blossom eventually. If there are no lotus buds nearby, you can place the stone in a glass jar, and a real ruby will emit a red ray of light. If that test is inconclusive and you happen to have a natural pearl on hand, put the pearl and ruby together on a silver plate in the sunlight. The silver will look black, the pearl will glow, and the ruby will shine like sun. This test isn't very practical, since natural pearls are harder to find than real rubies these days.

There are three basic ways of "counterfeiting" rubies. The first is to present a different mineral, like a spinel, as a real ruby. This was not always done on purpose. In fact, until the beginning of the nineteenth century, nearly all red gems were called rubies automatically. In 315 BC the Greek writer Theophrastus conveniently categorized all gems by color, so that garnets, rubies, and spinels were all the same. He called them all "carbuncles," a term which refers to the colors of glowing embers. However, it is also true that the gem trade honors several false stones with the royal name of ruby in an undoubted attempt to gain sales.

The humble pyrope garnet, for example (the dark red stone so popular in the late nineteenth century), is variously known as the Adelaide ruby, the American ruby, the Arizona ruby, the Australian ruby, the Bohemian ruby, the California ruby, the Cape ruby, the Colorado ruby, the Elie ruby, the Montana Ruby, and the Rocky Mountain ruby. The almandine garnet is sometimes called the Alabandine ruby. Rose quartz can masquerade as the Ancona Ruby. There's even a "garnet ruby," which of course is a garnet. Spinel passes for Balas ruby (its most ancient name), or the Ruby Spinel or Spinel Ruby. Cuprite is the Copper Ruby. Red tourmaline sometimes goes by the moniker of "Siberian ruby." And of course many synthetic rubies, such as the Geneva ruby, imply that they are the natural article. The oldest imitator of the ruby is simple glass. Glass however, feels warmer to the touch than ruby, and because it is so soft, usually is chipped and worn. Simply peering through a jeweler's loupe at a glass "ruby" will reveal swirls, bubbles, and other characteristics that are not at all typical of ruby.

There is a distinct difference between an imitation ruby and a synthetic ruby. A synthetic ruby is of the same chemical composition as a natural found-in-the-ground ruby. An imitation ruby is another substance, such as glass or another less-valuable gemstone that is called a ruby. If the seller informs the buyer what the product really is, there's nothing wrong with either practice. Only when the imitation or synthetic ruby is sold as a natural *and* genuine ruby is there a problem. (A synthetic ruby is genuine; it is not natural. A "Colorado ruby" is natural, but it is not a genuine ruby. A piece of colored glass is neither.)

Recently shady dealers in Asia have produced "doublets" to further confuse the buyer. A doublet consists of two parts, as the name suggests. The top part of the stone is simply cheap yellow or green corundum with all the natural inclusion typical of that mineral species, while the lower part is synthetic ruby, stuck onto the corundum by transparent cement.

Another way to enhance the appearance of a stone is by heat-treating or otherwise altering a natural ruby to improve its color. Nowadays nearly all rubies are heat-treated, and it is not generally regarded as cheating, unless the seller pretends the stone has not in fact been heated. A third way is the making of synthetic rubies. Again, there is nothing wrong with this so long as the buyer is aware that the stone is synthetic and not natural. We'll look at each of these in turn. But first we should take alook at one other way to trick the public.

THE GREAT DIAMOND (AND RUBY) HOAX OF 1872

I have to be honest. There aren't any ruby mines in Wyoming. But back in 1872, a fabulous diamond and ruby mine was "discovered," glittering away in Rock Creek, Wyoming. The developers, experienced prospectors Philip Arnold, J.B. Cooper, and Edward Slack, managed to collect hundreds of thousands of dollars from gullible (and greedy) eastern investors, among whom were Horace Greeley and even the jeweler Charles Tiffany. In point of fact, Tiffany knew almost nothing about diamonds, a fact he kept secret. He was actually a silversmith by trade, and had never done an apprenticeship in one of the great diamond houses of Europe.

Unfortunately for the criminals, Yale geologist and later director of the U.S. Geological Survey Clarence King investigated the mine. King has just done a geological survey of the 40th Parallel and couldn't understand how he had missed finding this incredible lode of diamonds. Suspicious of the unnaturally regular ratio between the two gems (12 rubies for every one diamond) everywhere he looked, as well as the peculiar placement of the diamonds, he proclaimed it a fraud. He wrote grimly, "Summing up the minerals, this rock has produced four distinct types of diamonds, Oriental rubies, garnets, spinels, sapphires, emeralds and amethysts—an association of minerals I believe of impossible occurrence in Nature." He was right. Arnold and Slack had bought 35,000 stones and salted the mine. Well, it almost worked. The whole deal was properly named the Great Diamond Fraud of 1872, but rubies were definitely involved.

SPINELS AS RUBIES: THE BLACK PRINCE, THE TIMUR RUBY, AND MORE

Only comparatively recently has it been possible to distinguish ruby from its many counterfeits. Spinel, garnet, and latterly the synthetic ruby, have all been mistaken for the genuine, natural article. Gem treatment and even fraud is nothing new. The ancient Roman historian and naturalist Pliny (23–79 CE), in his modestly titled *History of the World*, wrote:

> ...Moreover, I have in my library certain books by authors now living, whom I would under no circumstances name, wherein there are descriptions as to how to give the color of emerald to rock crystal and how to imitate other transparent gems. ...To tell the truth, there is no fraud or deceit in the world which yields greater gain and profit than that of counterfeiting gems.

However, Pliny was no fool. About "carbunculi" (or "red stones" in general), he wrote:

> False carbunculi are detected by lack of hardness of their powder and by their weight ...further, one sees in false carbunculi certain small inclusion, that is, blisters and vesicules, that resemble silver.

Dark red spinel is ruby's great imitator. While ruby is aluminum oxide, spinel is aluminum magnesium oxide ($MgAl_2O_4$). Like ruby, it gets its red color from chromium, that most chameleon of metals. Not that a spinel is anything to disparage. In itself it is a rare gem. It has almost the same geological history and is often found along with rubies and sapphires. It has the same vitreous luster and close to the same color. It's no wonder the two were constantly being mixed up. Spinels also grow larger and tend to have fewer flaws than rubies.

But although it is so close, it is not ruby. That didn't stop people from calling spinels rubies anyway, either through ignorance, fraud, or because they simply didn't care. The darker shades were sometimes called "spinel rubies," the lighter ones "balas" rubies, the orange red variety (colored by iron) "rubicelles," and the "almandine" a sort of violet red (colored by manganese), but all were spinels nonetheless.

Almost all "rubies" boasting hundreds of carats, including those in the French and Russian crowns as well as in the Iranian treasury, are actually spinels. Spinel has a hardness of 8, about the same as topaz. Its structure is basically octahedral, like a diamond. At one time it was known as the "Balas ruby," and is found in many of the same places that ruby is, including Myanmar and Sri Lanka. Although most large spinels today come from Burma, this may not have always been the case.

The Black Prince's Ruby

One of the world's most famous rubies, in fact, is not a ruby at all but a spinel. Its source is unknown; most experts guess Badakshan or somewhere else in Burma or Thailand, if only because these places were the only known sources for such stones at the time. This is the Black Prince's Ruby, two inches long and estimated to weigh somewhere between 160 and 170 carats. It has belonged to English royalty since 1367. The stone is now mounted in front of the Imperial State Crown (designed for Queen Victoria's coronation), above the 317.4 carat "Cullinan Two," the world's second largest diamond. It is semi-polished, and drilled in one end. The hole is filled with a real ruby, and the whole stone is fitted with gold-foil backing. The piercing of the stone has caused some controversy, and it is uncertain as to whether the piercing was done in England for the purpose of mounting a cross perpendicularly over it or whether it was done somewhere in the Middle East. (One source, William Fernie, maintains that the hole was drilled to hold the Black Prince's feathers.)

The stone was officially first recorded at the coronation of William and Mary of Orange, where it was called "the King's Great Ruby." But its actual history is much older, and is a matter of great controversy among gem historians.

It is agreed that the stone is named after Edward of Woodstock, the so-called Black Prince (1330–1376), eldest son of King Edward III and father of the future Richard II. No one has the remotest idea how he got the name "Black Prince," and he was not known by that epithet during his lifetime. Perhaps the name came from his documented cruelties. His recorded activities include the slaughter of 3,000 civilian inhabitants of the French city of Limoges in 1370. Others suggest that the name may derive from a suit of black armor he may or may not have actually worn. Although the son and the father of a king of England, Edward never got to rule himself, as he died a year before his father, of some sort of swelling of the lymph nodes.

In his life, the Black Prince is known for capturing John II ("the Good") of France, who reigned 1350–1364, at Poitiers. (The Encyclopedia Britannica dismisses John the Good as a "mediocrity.") John the Good was also mixed up with rubies. According to the Italian Renaissance poet Petrarch, he constantly wore a ruby cluster ring for good luck. Obviously this didn't work out as well as it might, as he let himself be taken prisoner at the Battle of Poitiers in 1356 by the Black Prince. (This was year before the Black Prince managed to get his hands on the "ruby" that now bears his name.) However, perhaps John's ring was somewhat effective after all, since the Black Prince treated him remarkably well and even prayed with him at Canterbury Cathedral.

Edward got the gem we call the Black Prince's Ruby from Don Pedro I, the King of Castile and Leon. Pedro (1334–1369) was more familiarly known as Pedro the Cruel, and so was a fitting companion for the Black Prince. The latter received the gem as a reward for helping Pedro defeat his brother Henry, King of Granada, in the battle of Najera in 1367, after Pedro had overrun his kingdom.

How Pedro got the gem is an interesting tale in itself. He is supposed to have taken it from Abu Said, the Moorish ruler of Granada. The story is that Pedro heard about this spectacular gem and decided the best way to obtain it was simply to invite Abu Said to his court in Castile for a little friendly get-together. Abu Said fell for the ruse, and arrived at the kingdom with his whole entourage. Pedro, of course, simply had him killed and kept the ruby. If any of this story is true, it seems to have made little sense for Pedro to hand over his prize to the Black Prince. More reliable history assures us that the Black Prince forced Pedro to give him the stone.

One of more bizarre tales about Pedro concerns his second wife Inez de Castro, cousin to his first deceased wife, who he married while he was still a prince. The marriage was strongly disapproved of by Pedro's father, Alphonso IV, who feared that if Inez bore a son, there might be all sorts of dynastic difficulties, so Alphonso IV took it upon himself to order Inez executed. Pedro was briefly upset by all this, but soon reconciled with his father. However, the story goes that when Pedro became king in his own right, he had Inez's body exhumed and seated on a throne. At any rate, Pedro was murdered on March 23, 1369.

The Black Prince's Ruby was officially added to the Crown Jewels in 1377 just in time for the coronation of Richard II (1367–1400). Richard became king at the age of 10 and was deposed in 1399 by Henry Bolingbroke (Henry IV).

The ruby was worn by King Henry V in his helmet as he led his small English army to victory against the French at the fateful battle of Agincourt in 1415. He credited the stone with warding off a blow meant for his head. It was a fairly important battle, with the outnumbered British coming through despite being rather severe underdogs. Shakespeare gave Henry V one of his very best speeches to honor the occasion. There is also a great song (Song on the Victory at Agincourt), written shortly after the battle. The stone was later reset and adorned many another Tudor and Stuart.

Richard III was supposed to have sported the stone during the Wars of the Roses at the Battle of Bosworth Field (1485). However, he wasn't as lucky with it as was Henry V, since Henry Tudor (later Henry VII) killed him there. In point of sad historical fact, at least according to Lord Twining, who examined the history of the Black Prince with some care, it is exceedingly unlikely that stone was actually worn at either battle.

In 1649, after Charles I (1600–1649) was beheaded, the gem disappeared from view, and some say that the great gem was sold for a mere 4 pounds and 11 shillings. If true, it miraculously reappeared at the Restoration in the Crown Jewels of James II, as the dismal period of Puritan rule in England came to an end. It was mounted in the front of the state crown (not on the top, as it had been in Edward VI's crown). This meant the hole in it had to be hidden, which is how the ruby came to be there.

The crown was redesigned again for Queen Victoria, made to be worn not just for her coronation but also for the opening of Parliament and for other state occasions. The Black Prince's Ruby was again in the front and set in a cross with the Stuart Sapphire beneath it in a circlet. However, when the Cullinan Diamond was cut, the Cullinan II (Second Star of Africa and second largest diamond in the world) replaced the Stuart Sapphire in the center of the circlet, which was retired to the back. (At one time the Stuart Sapphire was on the front of the crown and the Black Prince's Ruby at the back. Tastes change.)

Today the royal crown resides in the Tower of London, but at one time it was kept in the Abbey Church of Westminster. Apparently, this was wasn't the safest spot, and it is said that some of the more valuable gems were plucked out and replaced by glass. The crown itself was in such bad shape that when James II came to the throne in 1685, it was so battered and denuded of jewelry that repairs and replacements had to be made at a cost of 12,000 pounds.

Not even the Tower of London was foolproof, as is evidenced by the famous near-theft of the Crown Jewels by the Irish Colonel Thomas Blood (1618–1680) in 1671. Blood got off to a good start in the world of crime, as he twice attempted to kidnap James Butler, the first Duke of Ormonde. Blood blamed him for confiscating his Irish landholdings. Both attempts failed, but instead of giving up, Butler turned his hand to jewels. He pretended to be a genteel parson and befriended Talbot Edwards, the 77-year-old keeper of the jewels at the Tower. (Obviously security wasn't a high priority.) He managed to convince the luckless Edwards to show the gems to his "nephew" and two of his "friends," then hit him on the head with a mallet, gagged him, and stabbed him in the stomach. (Or perhaps one of the other conspirators did the stabbing. The evidence is murky.)

Blood then used the convenient mallet to flatten out St. Edward's crown and stuck it under a loose-fitting clerical robe. He stuck the Black Prince Ruby in his pocket. Unfortunately for him, Edwards's son picked that moment to visit his father, not having done so in years. He sounded the alarm and all the conspirators were arrested. (The senior Edwards had also managed to get his gag off.)

During his trial, Blood kept insisting that he wouldn't talk to anyone but the king, and unbelievably, the king humored him (after the trial at which Blood was not surprisingly sentenced to death). The king (Charles II) then asked him, "What would you do if I gave you your life?" Blood responded, "I should try to deserve it, sire." And the king pardoned him not only for that crime but for any crimes previously committed. He was restored to his lands in Ireland (which had been lost in the Restoration) as well as the annuity that went with them. Sometimes there is no understanding royalty. They live in a world of their own.

Despite the pardon, Blood was thoroughly distrusted. The diarist John Evelyn, who met him at social events, described him as "an impudent bold fellow" with "a villainous unmerciful look, a false countenance, but very well spoke and dangerously insinuating." At any rate, Blood was not much mourned, as his epitaph shows:

> Here lies the man who boldly hath run through
> More villanies than England ever knew;
> And ne're to any friend he had was true.
> Here let him then by all unpitied lie,
> And let's rejoice his time was come to die.

Nowadays the Imperial State Crown is trotted out for special occasions like the coronation of a new monarch or during important state occasions. There is actually another coronation crown, St. Edward's Crown, which is used for most of the coronation ceremony. The Imperial Crown is bigger and has more jewels on it. It is used after the actual coronation, when the monarch leaves Westminster Abbey, and also for the opening of Parliament. When it is actually on the Monarch's head, and thus not in the museum to be examined by tourists, a placard is put in its place that announces "In Use." In addition to the Black Prince's Ruby, it features more than 2,800 diamonds, including the Cullinan II. The sapphire in the Maltese Cross (at the top of the crown) allegedly comes from the ring of King Edward the Confessor. The 104-carat Stuart sapphire is set into the rim, opposite the Black Prince's Ruby. It was later discovered that the Black Prince's Ruby was the Black Prince's Spinel.

Its historic value far surpasses its gemological worth; at least that is the assumption. However, no contemporary gemological reports have ever been done of the Black Prince. This is too bad; there is probably much to be learned from it.

Henry VIII's "Ruby" Collar and the Three Brethren

Henry VIII owned two "collars" of balas rubies, which were really made from spinels. He wore the most important one, which consisted of 13 stones, some square, some oval. All of them were large and linked together by foliated gold ornaments, each set with sixteen pearls at his coronation.

Another collar was that of the famous Three Brethren. This is the collar that was probably later broken up by Charles I and sold to raise funds for his army. This may have brought him bad luck, as he was eventually beheaded. At any rate, it is all part of the strange and sordid tale of the "Three Brethren," three very large, rectangular, perfectly matched balas rubies or spinels. They are spaced by three big (very big) pearls. The three rubies appear in a pendant that features a point-cut (pyramid shaped) diamond created by

Louis de Berquen (fl. 1450–1470), probably the best known diamond cutter of the day. The Three Brethren themselves (without the diamonds and pearls) first turned up in 1419 inventory.

As with most jewelry of this period there are conflicting stories about provenance and who bought it from whom and when. One story claims that Charles the Bold (1433–1477) inherited it, but lost it at the Battle of Granson against the Swiss in 1476. It is said that a soldier looted his tent and took the jewels. Precisely the same story was told (inaccurately) about the Sancy Diamond. Both tales may be made up. We do know that Jacob Fugger bought it, and his son tried to sell it to Henry VIII just before that monarch died in 1547. Henry's son Edward VI did buy it, and it was later given to his sister Mary I (Bloody Mary) upon the occasion of her marriage to the Spanish Philip II. Queen Elizabeth I also wore it. It was reset and was worn by the future Charles I on his luckless visit to Spain, in which he attempted to marry the Infanta in 1623.

In 1626 Charles I pawned the Three Brethren in the Netherlands and redeemed them six years later. Charles I sent his wife, Henrietta Maria, to pawn them again in 1642, in order to help fund the English Civil War. This time they were never recovered.

The Samarian Spinel

The world's largest spinel is the blood-red Samarian Spinel, part of the Iranian Crown Jewel Collection. It is 5.5 cm (2.17 in.) wide and weighs 500 carats. It has a hole in it, although it is now plugged up. According to a diary entry of the court physician to Nasseridin Shah, the gem had once adorned the neck of the Golden Calf (Exodus), and a diamond later covered the hole. If there ever was a diamond covering the hole, it's gone now. As the for the Golden Calf story, it is probably just that—a story.

A companion stone, weighing 270 carats, is engraved with the name of the Mughal emperor Jahangir, so it may have more historical importance. The Mughal emperors were quite fond of engraving their names in gems, a rather expensive sort of graffiti. At any rate, both stones are believed to come from the balas ruby mines of Badakshan (modern Afghanistan and Uzbekistan).

The Russian Imperial Crown "Ruby"

Many of the great spinels are in the Kremlin's famous "Diamond Fund," the first version of which was instituted by Peter I (reigned 1683–1725), and which is really the Russian version of the crown jewels, gems owned by the state and not by particular members of the royal family. Most of the pieces have now been sold off, although a significant minority is available to the viewing public at the state armory in the Kremlin.

One 100 carat spinel is probably Sri Lankan, but most of the rest are from the mysterious Badakhshan. One is a glorious 398.72 carat stone that adorns the Russian "Great Imperial Crown." This stone is one of the largest spinels known. The crown was created by the court jeweler Jeremia Posier (Pauzie) for the Empress Catherine the Great's Coronation in 1762. It is made of two open hemispheres divided by a foliate garland and fastened with a hoop. This is a classical design inherited from the Roman Empire—the separate hemispheres representing the eastern and western parts of the empire. The crown is ornamented with about 5,000 Indian diamonds and almost the same number of rubies as well as a number of large white pearls. The "crown of the crown," however, is the big spinel that was carried to Russia by Nicolai Spafary, the Russian envoy to China from 1675–1678. Atop the spinel is a jeweled cross, which represents the Christian faith of the Czar. Unfortunately, the crown wasn't actually finished in time for the big day, but it was kept around until 1917 and the Revolution.

According to legend, the crowning spinel was found by a certain Chun Li, a Chinese mercenary and member of Tamerlane's (Timur's) army when it looted Samarkand. Chun Li decided that he would keep some of the booty (which did not include the famous spinel) for himself. He was found out and exiled to slavery in the Badakhshan mines, where he was lucky enough to find the stone and make away with it in the dead of night. His plan was to present it to the Chinese emperor in return for some massive reward, but unfortunately, he was murdered by a palace guard, who scooped up the gem for himself. The spinel must have been an exceedingly unlucky stone, however, since the guard himself was then killed by an informant-jeweler when he tried to pass it off. The Chinese emperor thus got the stone anyway. How much of this tale is true is anyone's guess. In 1676, Nikolai Spafari bought it from the Emperor Kon Khan on behalf of Czar Alexei Mikhailovich, the second Romanov Czar.

Another Russian Gem: The Zubov Ruby/Spinel

Plato Zubov was a young guard in the service of the Russian Empress Catherine the Great. He was the last of her lovers (he became so when he was 22 and she was 60) and remained with her until the day she died, which, contrary to anything you may have heard, occurred of a stroke in her water closet in 1796. She was 67 years old.

The Zubov family acquired a fine stone (presumably a royal gift) that went on loan to London's Victoria and Albert Museum, known as the Zubov Ruby. The Museum, however, using documentation from the family, determined it was a spinel, and labeled it as such. However, someone pulled a switcheroo, because when famed gemologist Fred Ward examined the stone carefully, he announced it was ruby after all—but not a natural one. In fact, it was a twentieth century flame-fusion synthetic. The original stone was probably a high-quality spinel, but somehow, somewhere, someone made the change. The original is lost, presumably forever.

The Timur "Ruby"

The Timur Ruby weighs in at 352.2 carats and is about two inches square. It is also known as Khiraj-i-alam ("Tribute to the World"), an inscription which Nadir Shah had engraved on it. The gem is semi-polished and engraved with the names of its former owners. Its source is uncertain, but is possibly from Badakshan. As is the case with the Black Prince stone, we unfortunately have no up-to-date gemological report on this stone. What is certain about this stone is that its fate is entwined with that of the fabulous Koh-i-Noor Diamond. Since 1612 the gems have always been owned by the same person, whether they were in India, Persia, Afghanistan, or England. Both the Kor-i-noor diamond and the Timur were once owned by Maharajah Ranjit Singh, for example, who was also charmingly known as the Lion of Lahore.

Until 1851 this stone was thought to be a ruby. Its size, 352.50 carats, puts it second behind the 398.72 carat spinel in the imperial Russian crown, mentioned earlier, and which was, of course, also once thought to be a ruby. Today any "ruby" of this size would arouse immediate suspicion in the minds of those who "know." The Timur ruby retains its original baroque shape, as does the smaller Black Prince's Ruby.

The stone is named after Timur (1336–1405), also known as Quiran and Tamerlane (in the West), who was one of the conquering Tartar "nomad-kings." Christopher Marlowe wrote a play about him, one of his worst. Tamerlane (Timur-i-Leng), as his name seems to suggest, really was lame, having sustained some sort of injury as a child.

Tamerlane is said to have come into the stone's possession in 1398, when he conquered and looted Delhi. At least, that was the traditional (but apparently mistaken) interpretation of the inscription written on the stone.

Nowadays, the association with Tamerlane is deemed spurious (something was mistranslated somewhere), but the name has stuck.

Tamerlane spent most of his time conquering the world, and went through India in 1398–1399. He wasn't very pleasant when doing his job. When he conquered Baghdad, for instance, he commanded every soldier to show up with a minimum of two severed human heads to show him. The soldier got around the decree simply by beheading the prisoners they had already captured so they wouldn't have to fight anyone else. After the death of Tamerlane, the stone came into the possession of Mir Shah Rukh, his third son and successor. Tamerlane's body was exhumed in 1941 by Russian anthropologist Mikhail Gerasimov, who thought that Tamerlane's facial features as reconstructed from his skull bore a resemblance to those of Genghis Khan, form whom Tamerlane had always claimed he was descended. There's a famous curse attached to the opening of the tomb, which stated that anyone who violated it would have his country attacked by demons of war. Indeed, Russia was invaded three days later by Germany.

The Timur Ruby is large enough to bear several inscriptions upon its face, in the Persian language but in Arabic script. The longest of them reads, "This ruby is among the 25,000 jewels of the King of Kings, the Sultan Sahib Qiran which in the year 1153 from the jewels of Hindustan reached this place." "This place" means Isfahan, the old capital of Persia, to which Nadir Shah brought it after he stole it.

There are five other inscriptions, which include the name of the various emperors who have owned it. The other "owners" of the stone were recorded on it as Akbar Shah (1021/1612), the Moghul Emperor of India; Jahangir Shah, Sahib Qiran Shani, otherwise known as Timur or Tamerlane (1038/1628); Alamgir Shah (1070/1659); Bagshesha Ghazi Muhammad Farukh Siyar (1125/1713);and Ahmad Shan Dur-i-Duran (1168/1754). The dates were originally inscribed in the Islamic calendar; the second date reflects the Western calendar. However, these men were not the only owners of the great spinel, and Akbar Shah, Jahangir's father, didn't own it at all—his named was carved there apparently by Jahangir as a filial courtesy.

Shah Jehan also built the famous Peacock Throne (supposed to have been initiated by Tamerlane) in Delhi and ornamented it with a multitude of gems, including the Timur Ruby. Supposedly the cost of the Peacock Throne was twice that of the Taj Mahal, although that seems hard to believe. It was a spectacular piece of furniture, by all accounts. The name comes from the fact that it has two peacocks standing behind it, with their expanded tails inlaid with rubies, sapphires, emeralds, pearls, and other gems of symbolic colors. Tavernier saw that throne or a similar one, but claimed it was in the shape of a columned bed adorned with rubies, emeralds, and exceedingly fine pearls. Some have estimated that its value today would be over a billion dollars, but that seems an exaggeration. In any case, in 1739 the throne was carted off to Persia by Nadir Shah and later destroyed. Later, the term "Peacock Throne" came to represent the Iranian Monarchy, which was quite curious, considering that the throne itself was of Indian manufacture.

At any rate, the Mir Shah Rukh who inherited the Timur spinel was not the same Mir Shah Rukh who got hold of the Koh-i-noor diamond (and who was tortured to death for it about 250 years later). It is not always easy to sort these people out. Mir Shah Rukh passed the stone along to his own son, Mirza Ulugh Begh (1394–1449), who was famous as an astronomer, publishing a famous star catalogue. There is a crater of the moon named after him. He was eventually beheaded by his eldest son, who was apparently quite irritated by his father's lack of administrative skills (having lost some important battles and massacring most of the people of Heart).

The Timur changed hands a number of times after this and eventually came into the possession of Shah Abbas Safari, who ruled India from 1587 to 1629. Mir Shah Rukh, Mirza Ulguh Begh, and Shah Abbas Safari all had their own names engraved on this very large stone. These engravings were later removed under the orders of Jahangir or another Mogul emperor. Jahangir, who ruled for 23 years, gave himself the name Jahangir, which means "conqueror of the world." Jahangir got the Timur as a present (he claimed) from Shah Abbas Safari. There were apparently some complaints about the damage done to the stone by all this engraving and rubbing out, but Jahangir is reputed to have stated, "This jewel will assuredly hand down my name to posterity more than any written history. The house of Timur may fall, but as long as there is a king, this jewel will have its price." It is true that Tamerlane's dynasty died out after 150 years, while Jehangir's name is still scratched on quite a number of gems now in the Iranian treasury and in museums around the world. Still, more people have heard of Tamerlane than of Jehangir, although quite an astonishing number of people have heard of neither. Rather unbelievably for emperors in those times, Jahangir died in his sleep.

Jahangir passed the stone on to his own son Shah Jahan (ruled 1630–1653) who had it from 1628-1658. Shah Jahan called himself the Sahib Quiran Sani, which means Second Lord of the Conjunction. (The first Lord of the Conjunction was Timur and the reference is to a certain conjunction of stars which was thought tremendously important at the time.)

Shah Jahan is best known for the great love he bore for his beautiful wife Mumtaz, whom he married in 1615, and who was of Persian, not Indian, background. They had 14 children together, and she shared in all his military campaigns. She died 1631, undoubtedly from exhaustion. But she got a very nice tomb—the Taj Mahal.

Shah Jahan had the Timur mounted in the Peacock Throne. Some of the world's most famous gemstones once adorned this cherished possession of the Mogul emperors. According to Tavernier, who claimed to have inspected it personally, the throne contained 108 balas rubies (probably spinels), all cut in cabochon, the least of which weighed 100 carats, with the largest at twice that weight.

The next name inscribed on the stone is that of Alamgir Shah, who is better known as Aurangzeb ("ornament of the throne") and ruled from 1658–1707. He is the one whose name is also recorded in relation to the Chhatrapati Manik ruby, since disappeared. Aurangzeb, in time-honored custom, imprisoned Shah Jahan, his father, in a fortress at Agra. He also murdered two of his own brothers. There is a persistent rumor that old Jahan placed a ruby with a convex surface into the wall of his prison so that he could see a miniature "mirror image" of his Taj Mahal in the distance. Rubies make remarkably poor mirrors, but it is a nice story all the same. Tavernier mentioned that Aurangzeb's uncle, Ja'fan Khan, presented him with a fine large stone, reputedly a ruby worth 95,000 rupees. However, during the presentation ceremony, an old court jeweler who had lost his job examined the stone and declared it was not a ruby at all. Unconvinced, the Emperor sent the stone to his father, the former emperor Shah Jahan, whom he had had locked up, and asked his opinion. Shah Jahan agreed with the former court jeweler, so Aurangzeb forced the merchant to take the stone back and refund the money. Unfortunately for Shah Jahan, this did not get him out of jail. The next two Moghul emperors did not engrave their names on the stone, but two later possessors did so.

In 1739, the Shah of Persia, Nadir Shah (1688–1747) invaded India, partly to seize this very stone, at the time believed to be the largest ruby in the world. It was Nadir Shah who ordered the long inscription engraved on the gem. The date is based on the Islamic calendar, which begins from the flight (Hegira) of Muhammad from Mecca to Medina, and corresponds to the western year of 1740.

Nadir Shah captured Delhi and took the Timur Ruby (and the Peacock Throne) back with him to Persia. After Shah's violent death in 1747, the stone went to Ahmad Shah Durani (Taimur Shah). The final Persian owner was Shah Shujah, who was imprisoned first in Kashmir and then in Lahore, where the stone came into the hands of Ranjit Singh, the Lion of Lahore. It then went to his successor Dhulip Singh. Not directly, however. One of

Ranjit Singh's sons took over but was deposed by his own son, who was in turn murdered and succeeded by his mother. In fact, so many people fought over the throne that eventually only one claimant was left standing, a child of a dancer and a water-hauler. A member of the harem had somehow managed to convince everyone that the child was Ranjit's.

Nadir Shah ruled Persia from 1736 to 1747. He was a fairly rough customer, by any account. He began life as a shepherd, but because of his imposing size and strength, soon became a bodyguard for a governor of Khorasan. Eventually he married the governor's daughter and tried to take over the region. He failed in the attempt and so decided to become a bandit instead, a profession at which he excelled. His fame grew to the point where the current Shah, Tahmasp, who was then without a kingdom, asked Nadir to help him regain power. Nadir obliged by simply removing Tahmasp completely and declaring himself Shah. (For a while he contented himself with being merely Regent for Tahmasp's eight-month-old son, but that sort of thing never lasts.) Nadir Shah was well-known for his severity with his generals. If one of them retreated from battle, Nadir simply dispatched him with his battle axe and gave the command to the next senior officer. In this way he kept his army moving steadily forward into India. Not surprisingly, Nadir Shah came to a rather bad end. He became ill with dropsy, went mad, blinded his eldest son during a temper tantrum, and was eventually murdered in his tent. No tears were shed.

When the British annexed the Punjab area in 1849 after the Sikh Revolt, they also acquired the Koh-i-Noor diamond and the Timur Ruby from the Lion of Lahore. It is said that earlier Baron Charles von Hugel, visiting the Maharajah in 1836, noticed this vast "ruby" inserted into the pommel of the Maharajah's saddle, and told everyone he knew.

The East India Company took possession of the stones and sent them to England as a present to Queen Victoria (1819–1901) in 1851, in thanks for her patronage of the Great Exhibition. In her diary, she wrote, "They are cabochons, uncut, unset, but pierced. The one is the largest in the world, therefore even more remarkable than the Koh-i-noor!" (October 23, 1851). Of course, she owned that stone as well.

In April of 1853 R & S Garrards set four of these so-called rubies in a diamond-encrusted gold and enamel necklace "of Oriental design," along with four diamond pendants also from Lahore. At the center of the necklace was the massive rose-pink stone that impressed Queen Victoria. Two months later, they redesigned the necklace to allow this stone to be detached and used as a brooch to alternate with the Koh-i-Noor.

It was listed in the official catalogue as a "Short necklace of four very large spinelle rubies." Queen Victoria wore the necklace occasionally, notably during the State Visit of Napoleon III and the Empress Eugénie in 1855. Queen Mary, the wife of George V and mother of Edward VIII who abdicated the throne, had the necklace lengthened in 1911, but it has seldom been worn since. Despite the Queen's pleasure in the stones, for 60 years the Timur was put into storage with three other spinels, ignored and forgotten.

Only when the inscriptions were examined was it realized what a treasure they had. Most of the attention had actually gone to its companion stone, the Koh-in-Noor diamond. Interestingly, the Koh-i-Noor and the Timur, no matter how much they have been fought over, have remained together since 1612. That is something of a record.

Today the Timur Ruby is in the private collection of Queen Elizabeth II, as she inherited it. It is not a crown jewel, although there are certainly those who believe that it should be. The Queen has also inherited Queen Mary's large ruby earrings. The ruby in each earring was set in a nest of nine brilliant diamonds. Queen Mary received them from her George V on May 26, 1926, in honor of her 59th birthday. The Queen also has a tiara that she actually commissioned (from Garrards) in 1973 rather than inherited. It is a stunning wreath of flowers made from Burmese rubies; there are diamonds too. A queen can never have too many tiaras, it seems. The rubies for the tiara came from another part of her collection, a wedding gift to her Majesty by the people of Burma. It was a necklace of 96 rubies set in gold. The number 96 has an interesting relevance; according to Burmese lore, the human body is prey to 96 diseases, and each of the rubies was a charm against one of the illnesses. Whether the charm still worked once the wedding present was broken up is not known, although it does seem a slight against the original givers.

The "Ruby" at the Kaaba

The Ka'aba at Mecca holds a seven-inch "ruby" revered as the stone of the Last Judgment. It is supposed to have ears, eyes, and a tongue, so that it can tell everything it knows when the fateful day arrives. Presently it resides in the northeast corner of the shrine and is now black from being kissed so many times by the sinful lips of humanity. Originally it was red (representing the female principle), along with a white stone (representing the male principle). However, Arabic legend tells us that the stone was originally a ruby sent down from heaven during the time of Adam and Eve. For unclear reasons, God deprived the stone of its amazing brilliance, which otherwise would have illuminated the universe from end to end. A similar legend says that Abraham brought the ruby with him when he set up the shrine. (He got it from Gabriel.)

The stone itself is pre-Islamic, having been worshipped for centuries before the arrival of Muhammad. It is probably a meteorite, as these objects were worshipped rather regularly in pre-Islamic Arabia. No one can be certain, however, as its guardians will not allow it to be removed and examined scientifically. However, it is fairly certain that whatever it is, it is not a ruby. The stone was actually stolen and broken up in 930 CE, but was returned 22 years later, mended with a silver band and nailed in its place with silver nails.

TREATING RUBIES: ENHANCING THE FIRE

The Roman naturalist and historian Pliny wrote: "To tell the truth, there is no fraud or deceit in the world which yields greater gain and profit than that of counterfeiting gems." He went on to talk about how gems were heated to improve their color, or even in some cases enhanced by the addition of foil as a backing.

A perfect natural gem is extraordinarily rare. People have been heating gemstones for at least 4,000 years to make them look better. Heat treatment has two effects: dissolving the rutile crystals and improving the color. Only stones (both rubies and sapphires) of exceptionally good color are left in their natural state—and also those stones deemed too far gone to be helped by heat treatment. Heat treatments can be generally detected only in gems with residual inclusions, which may show signs of heat stress, although heat treatment does not remove flaws. Clean stones (very rare) may show evidence of heat treatment. However, many heat-treated rubies (and sapphires) will display chalky short-wave fluorescence never found in untreated corundum. Weirdly, it is the colorless parts of the stones that actually fluoresce in this case.

Even most rubies that are too dark can be heated in oxidizing conditions to improve their color. Unless the seller specifically affirms that the ruby is unheated, it is assumed that it has been heat-enhanced.

Usually high temperature heating in an oxidizing atmosphere and controlled cooling is done to clarify the stones, especially by dissolving "silk" (rutile crystals), but it can also improve tone and saturation of color by removing unwanted blue tints. In fact, between 90 and 95 percent of all rubies on the market, at least anything that has gone through Thailand, has undergone heat treatment, including nearly all stones from Mong Hsu, which are generally not facetable without it.

Too many rutile crystals give a ruby a "cloudy" and distinctly undesirable appearance. During the treatment, the "silk" dissolves back into the corundum without damaging the ruby, and the stone looks much clearer. In fact, the rutile needles create a slightly softer appearance, which is preferred by some buyers. Many merchants maintain that this simple heating is indistinguishable from the natural processes. Heat enhancement is stable, inexpensive, and permanent; the rutile needles won't re-form afterwards.

What to call heat alteration is a bit of a problem. The American Gem Trade Association, whose members are wholesale jewelers, differentiates between "treatment" and "enhancement," preferring the latter term, although in the view of most people, there's really no difference. The Federal Trade Commission has produced guidelines requiring disclosure of any treatment that significantly affects a stone's value.

Before 1990, heat treatment was seldom done in Burma, partly because Mogok rough doesn't respond well to the crude furnaces available in that country, which could not raise temperatures sufficiently high.

In ancient times the heating was done with charcoal fires and blowpipes. Today, it is carried out in computer-controlled electric furnaces, which use very carefully calibrated heating/cooling sequences.

Heating a ruby properly is a delicate task, however. To improve the clarity, the stone is heated to between 1000 and 1900 degrees C or 1831 and 3452 degrees F, a point somewhere between the melting point of rutile and the melting point of ruby (about 2040 degrees C), for several hours and then cooled rapidly under very controlled conditions. Amazingly enough, native Sri Lankan ruby "cookers" can raise the temperature of a stone to above 1400 degrees C by puffing on a blowpipe twice per second. The traditional fuel is coconut husk charcoal, although more modern "facilities" use bottled gas. In Thailand, ruby treatment methods were long kept a secret; modern treatments make use of electric furnaces.

You can't wash away the new color, and soap doesn't hurt it a bit. Therefore, so long as the process is disclosed, heating does not *substantially* affect the value of stones. However, a glorious natural-colored ruby, if you can find one, is still worth more than a heat-treated one of the same shade and appearance. Star rubies are not treated, as it would destroy the rutile needles that cause the asterism.

In addition to heat treatment, occasionally you will come across a ruby that has been oiled or dyed. This treatment is usually done to lower-quality stones, and is not permanent. If the buyer is not told about a color treatment, it is considered a deceptive practice.

Some rubies, though, have more than a color problem. They have fissures that break through to the surface. These are sometimes filled with a glassy substance that's a by-product of the heating process. Others are filled with glass, solidified borax, or other materials to improve their looks and durability. These treatments are usually not so permanent as heat treatment and the stones may wear over time; the fillings can even fall out if the stone is roughly handled. While it is not unethical to sell such a stone, it is unethical to pretend that a heated or fracture-filled stone is natural. It is not. Stones treated in this way are worth much less than natural stones. Another problem with them is that if they are cleaned ultrasonically, a fracture-filled stone can be seriously damaged.

One of the latest treatment methods of rubies and other gems is called "lattice diffusion," a process developed by the Thais. In lattice diffusion, beryllium is diffused into gemstones by coating stones with a beryllium paste and then heating the stone to 1800 degrees C for hours or even days. The process works best on smaller stones; larger ones often acquire the color only about halfway through. The biggest beneficiary of the procedure is small colorless corundum crystals (of which the world has an ample supply). Suddenly, formerly valueless stones are appearing in shocking pink, gold, and forest green.

Through this process, too, the mediocre rubies from Tanzania's Mehenge mining district turned a brilliant, vibrant red. It is possible for a gem lab to detect lattice diffused stones, but the test itself is expensive—running to several hundred dollars per stone.

THE ALCHEMIST'S ART: THE SYNTHETIC RUBY AND THE LASER BEAM

The first method of making synthetic gems was simply to take fragments of the natural stone and fuse them together; the result was "reconstructed rubies." However, this was not particularly successful.

The first synthetic rubies appeared mysteriously in 1885, and would have perhaps gone undetected forever, except that the buyer was suspicious about their low price and had them checked. These first stones were known as "Geneva" rubies, but how they were created remained somewhat of a mystery until 1970. In that year, an analysis was made on the surviving samples, and it was clear they were made by melting powdered aluminum and chromium oxide. This was an early version of what is now known as the "flame fusion process." This process was invented by the French chemist Edmond Frémy some time before 1877. He and a student assistant heated 20–30 kilograms of a solution of aluminum oxide dissolved in lead oxide in a porcelain vat for 20 days. As the solvent evaporated, chemical reactions occurred in the solution, the porcelain vessel, and furnace gases, resulting in a great many very small ruby crystals forming on the wall of the porcelain. It was an undoubted success, but the rubies were so small and the production costs so high that the process was not practical for making gems.

In 1902, the process was improved by French chemist Auguste Verneuil (1856–1913), a student of Frémy. He had actually perfected the process years earlier and deposited his sealed notes at the Paris Academy of Science, but for reasons of his own, kept mum about it until 1902. His process took a mere two hours to grow rough crystals weighing 12–15 carats up to a quarter inch in diameter. By the time Verneuil died, his process was being used to manufacture 10 million carats (4,400 pounds) of rubies every year.

The flame fusion process works by exposing finely powdered aluminum oxide and a coloring matter (or doping agent) like chromium oxide to the flame of a blowtorch or furnace at around 3,600 degrees F. A hammer atop the apparatus strikes the hopper repeatedly, causing a small amount of the powder to fall through the fine mesh that forms the hopper's floor.

The material melts as it falls through flames and drops onto a "pedestal," which results in a shaped solid called a "boule." After about five and a half hours, the crystal reaches a length of approximately 2.75 inches. Then the gas flow is halted, extinguishing the flame. The crystal, now weighing around 150 carats, is allowed to cool in the enclosed furnace. The stone can then be faceted. The Linde Division of Union Carbine Corporation modified Verneuil's flame fusion process to grow thin rods of ruby crystals up to 30 inches long, which can be sliced into discs to produce large quantities of bearings for industrial use.

When first developed, synthetic rubies created a tremendous drop in ruby prices world-wide, especially since no one at the time had any way of telling the difference between natural and synthetic stones. It took several years for prices to stabilize.

Flame fusion is still the least expensive method for creating synthetic rubies. It is also very fast and can actually create a ruby within a few hours, as opposed to the several thousand years it took nature. (When you buy a natural ruby, you are therefore paying for the time put into the creating of it and the finding of it as well as the cutting of it.)

However, today it is obvious that flame-fusion rubies are not of the best quality. They can easily be detected by even a handheld loupe. The growth lines of corundum produced by this process are curved rather than straight, as the ingredients have not completely mixed. Some of them also have uneven color distribution. While this is also true of some natural rubies, it's definitely not desirable in either. (Synthetic spinel is made the same way, and then often colored blue to imitate aquamarine or other stones. Curiously, or perhaps not so, synthetic red spinel often makes a better natural ruby imitation than does flame-fusion corundum.)

Flame-fusion rubies are found today almost entirely in lower priced jewelry such as class rings and watch bearings. That is a story in itself. The idea for using jewels as watch bearings began in England, but it was the Swiss who in 1830 started creating natural ruby "doughnuts."

A third method of producing synthetic rubies is the Czochralski "pulled growth" or melt rubies, used mostly in high-tech applications. The method was developed in 1917. It's called "crystal pulling," and it's not only fast and cheap, but actually produces flawless, transparent stones. (They are so transparent that they look like glass). Here the ingredient powders or nutrients are melted in a crucible. A seed crystal is attached to one end of a rotating rod, which is lowered into the crucible until the seed just touches the melt. Then the rod is slowly withdrawn. The crystal grows as the seed "pulls" materials (hence the name) from the melt. The material then cools and solidifies. Typically, the seed is pulled from the melt at a rate of 1 to 100 millimeters per hour. Crystals grown using this method can become extremely large, more than 50 millimeters in diameter and 1 meter in length, and of high purity. Millions of ruby carats are grown this way every year, primarily for manufacturing industrial-use rubies, especially lasers. They can be bought as cheaply as five dollars per carat.

Both flame fusion and crystal pulling are varieties of what are called "melt techniques," where powdered material is simply heated to a molten state and manipulated to solidify in a crystalline form.

In 1958 Bell Telephone Company developed a process requiring both high temperature and pressure to grow rubies on seeds that had been produced by flame fusion, which (with some minor improvements) became known as the hydrothermal method. Here, the nutrient (either powdered or crystalline) is lodged at one end of a pressure-resistant tube. This tube can be sized to the desired dimensions. Then a seed crystal is mounted near the other end of the tube. An aqueous solution is placed in the tube, which is sealed shut and placed vertically in a furnace chamber. The lower, nutrient-containing end of the tube is heated to about 835 degrees F, compared to 770 degrees F for the top. Dissolved nutrient material migrates toward the seed, crystallizing on the relatively cooler surface. The method takes about a month, with crystals growing at a rate of about 0.006 inches per day during the 30-day processing period. Today this process, which is also used for making emeralds, is most used for industrial application rubies, which require strain-free stones or large crystals in something other than a rod shape.

Carroll Chatham of San Francisco developed the first commercially successful application of the so-called flux process in 1959 in his basement. It imitates nature by creating roiling magma in a furnace. In chemistry lingo, a flux is any material that when melted will dissolve another material that has a much higher melting point. Even though temperatures in excess of 3600 degrees F (2000 degrees C) are needed to melt ruby's main ingredient, aluminum oxide, the material will dissolve in certain fluxes at a temperature as low as 1470 degrees F. However, manufacturers prefer to use temperatures above 2200 degrees F, as they result in higher-quality crystals. The temperature is maintained for a period of three to 12 months.

During the process, the corundum molecules move freely, attaching themselves to a growing crystal. (Some manufacturers immerse seed crystals in the solution, while others take a more serendipitous attitude and let random formation occur.)

The process takes almost a year, but the result is an extraordinarily natural-appearing gem. The flux process produces rubies that cost $50 or more per carat, and are used in fine jewelry.

This method and the hydrothermal method are both called production from solution, where aluminum oxide and chromium are dissolved in another material and manipulated to precipitate into a crystalline form.

In fact, it's even possible to manufacture synthetic star stones by adding rutile (titanium oxide) to the feed powder. This produces rutile inclusions that form the rays in the synthetic star rubies. These synthetic star rubies (called Linde Stars) look better than the natural ones, since the rays tend to be straighter. The cost is a little higher than for a regular synthetic ruby, but nothing near what a natural star ruby would cost.

No matter what the process, the starting ("nutrient") materials are basically the same. The nutrient (material that will become the ruby crystal) consists primarily of extremely pure aluminum oxide; 5–8% of chromium oxide must be added to produce the essential red color. If a star ruby is the object, a small amount (0.1–0.5%) of titanium oxide is also used. Other chemicals may be needed as part of the process. For example, the flame fusion process requires an oxygen-hydrogen torch to melt powdered forms of the ingredients, while the Czochralski process uses an electrical heating mechanism instead.

The flux method uses a compound like lithium oxide, molybdenum oxide, or lead fluoride as a solvent for the nutrient. The hydrothermal process uses an aqueous solution of sodium carbonate as a solvent. Silver, platinum, or some similar corrosion-resistant metal lines the container that contains the liquefied ingredients for the Czochralski, flux, and hydrothermal processes.

Still another method used to grow rubies is the "heat-exchanger method" developed by Crystal System, Inc. This process is used for creating ruby lasers and has resulted in ruby crystals up to 65 kilograms in size.

No matter what method was used to create the synthetic ruby, it must be cut, faceted, and polished just like a natural ruby. In some cases, the stone is also "glossed" after initial polishing, which means heating it rapidly in a gas flame to melt any extraneous exuberances on the stone. It has the added benefit of nearly doubling the stone's tensile strength.

While it is fairly easy to distinguish between industrial-use synthetic rubies and the natural stone, it's not so simple a matter with synthetics intended for the fine gem trade. However, it *is* often possible to do so by carefully examining (with a microscope) the pattern of inclusions. Some manufactured rubies have "dopants" added to them so that they can be identified as synthetic; however, most require gemological testing to make that determination. Reputable dealers, of course, do not attempt to hide the fact that they are selling synthetic stones.

The production of synthetic gems has changed the face of the modern jewelry world—and of science. Synthetic rubies are not "fake" gems, since they have not only the same color, but also the same chemical composition and same optical and physical properties as do the natural gems. One of their great advantages is the variety of shapes into which they can be cut—not simply the classical oval shape of natural rubies, but also fancy shapes. Since synthetic stones can attain weights of 5–15 carats, and still remain astoundingly transparent, there is a lot more that can be done with them. They can even be beaded, something almost never done with fine natural rubies. Natural ruby beads are generally of poor quality, good rubies being too rare to be strung up in beads; they deserve to be shown off.

The hardness of ruby has given it some commercial importance. Ruby makes a long-lasting thread guide for textile machines, and makes an excellent bearing material for use in watches, compasses, and electric meters. Rubies also have extremely good wave-transmitting properties from short, ultraviolet wavelengths all the way through the visible light spectrum to long, infrared wavelengths. This makes them ideal for use in lasers and masers (laser-like devices operating in non-visible ranges of microwaves and radio waves).

Because many of these industrial uses demand very high-quality crystals of particular sizes and shapes, synthetic rubies are manufactured. Although some synthetic stones are used as gemstones, about 75 percent of modern synthetic ruby production is used for industrial purposes.

Rubies were the world's first synthetic gems to become available in large quantities, and are available for only a few dollars per carat. This produces consternation among gem dealers. One of the elements that make a stone valuable is its rarity. If synthetic stones, which are identical in every respect to flawless natural ones, can be conjured up cheaply, it's difficult to maintain the "value" of the natural stone. Some have suggested adding "natural" to the list of qualities that makes a stone precious; only time will tell if such a ploy works. It well may, since almost the entire value of gems are in the perception.

Currently the biggest producer of corundum is H. Djévahirdjian in Switzerland, which creates synthetic stones literally by the ton using the flame-fusion process. Most of its product goes into inexpensive jewelry and watch crystals.

The latest in synthetic rubies may be the "cultured" gems developed by Ramaura, a selection of which was donated to the Smithsonian's National Gem and Mineral Collection. The eight stones were created by Judith Osmer, CEO of the J.O. Crystal Company, Inc., of Los Angeles. These stones are created with a "self-nucleating process," a variety of flux growth, in which no "seeds" are needed to trigger the growth of the stone. These synthetic stones have the color and quality of the best Burmese rubies, but may grow into shapes and patterns rarely seen in nature. However, the object is to imitate, not to deceive. Every Ramaura ruby contains an identifier actually built into the chemistry of the stone to help gemologists aid in its identification as a synthetic ruby, not a natural ruby. Many even have inclusions. However, there's a catch. People who are willing to pay "ruby-prices" generally want natural rubies.

To make synthetic rubies even more natural looking, manufacturers sometimes produce internal fractures in them by first heating them and then cooling them rapidly. This practice makes good synthetics increasingly hard to tell from the natural stone. This is not done so much any more, as it looks suspiciously like "cheating."

Lasers

Rubies have their practical, as well as their beautiful, aspects. (Even the great ruby imitator, spinel, was once magnetized and used as a lodestone.) Primary among these is the use of the ruby for lasers, although I should say that natural rubies are of no use here—the key element chromium must be "doped up" in synthetic rubies to create the laser. This is just another reason why the synthetic development of rubies has had a practical effect on the contemporary world. The word "laser" stands for "light amplification by stimulated emission of radiation." Interestingly, the laser was invented before any real use was discovered for it. It was, in the words of Harry Stone, "an invention looking for a job."

The ruby laser was the first laser invented in 1960, although a bitter legal battle ensued as to who actually would get the credit (and the money). Credit now is usually given to Dr. T.H. Maiman. The key is chromium, that magic element that makes rubies red. Chromium atoms absorb green and blue light and emit or reflect only red light. Chromium is responsible for the "lasing" behavior of the crystal.

Lasers work by taking advantage of how light interacts with electrons. As we learned earlier, electrons exist at specific energy levels characteristic of each particular atom or molecule. You can envision these energy levels as rings around a nucleus, even though the reality is somewhat more complicated. Electrons in outer rings "buzz" at higher energy levels than those in inner rings. If you flash a light (which is energy) at an electron you can "bump it up" to a higher energy level. Likewise, when an electron drops from an outer to an inner level, its "excess" energy is given off as light. The wavelength or color of the emitted light is precisely related to the amount of energy the electron releases as it falls from a higher energy level to a lower one.

To make a ruby laser, a crystal of ruby is shaped into a cylinder. A fully reflecting mirror is placed at one end and a partially reflecting ("half-silvered") mirror at the other. A high-intensity lamp or flash tube is spiraled around the ruby cylinder to provide a flash of white light that triggers the laser action.

The green and blue wavelengths in the flash excite electrons in the chromium atoms into a higher energy state. As they return to their normal state, the electrons emit their characteristic red light (the fluorescence of natural rubies). The mirrors reflect some of this light back and forth inside the ruby crystal, stimulating other excited chromium atoms to produce more red light, until the light pulse builds up to high power and drains the energy stored in the crystal.

The laser flash that escapes through the partially reflecting mirror lasts for only about 300 millionths of a second, but it is extremely intense. Even early lasers could produce peak powers of some ten thousand watts. New ones can produce pulses that are billions of times more powerful.

Laser light is more than intense. It is also "coherent," which means that emitted light waves are "in phase" with one another and are so nearly parallel that they can travel for long distances without spreading out. Technically this is called monochromatic, single-phase, columnated light. Compare this to the "incoherent" light from a light bulb. This feature means that laser light can be aimed with extraordinarily precision. It turns out the ruby has an almost magical power after all. Lasers can even be used to drill out the defects in diamonds.

The ruby has traveled a long road, and has taken us with it. It has touched humankind in every conceivable way: spiritually, economically, aesthetically, technologically, and magically. We have kept company for a long time. But even after a long and intimate acquaintance, this stone is still a stranger in many ways. The stone still keeps its secrets. It won't easily surrender them.

Selected Bibliography

Field, Leslie. *The Jewels of Queen Elizabeth II: Her Personal Collection.* New York: Harry N. Abrams, Inc., 1987, 1992.

Gübelin, Eduard, and Franz Xaver-Erni. *Gemstones: Symbols of Beauty and Power.* Tucson, AZ: Geoscience Press, 2000.

Hughes, R.W. *Ruby & Sapphire.* Boulder, CO: RWH Publishing, 1997.

McCrane, A., and A. Prentice. *Irrawaddy Flotilla.* Paisley: James Paton, Limited, 1978.

Penzer, N.M. *The Mineralogical Resources of Burma.* London: George Routledge & Sons, Ltd; New York: E.P. Dutton & Co., 1922.

Peretti, Adi. "Rubies from Mong Hsu," *Gems and Gemology* (Spring, 1995).

Samuels, S.K. *Burma Ruby.* Tucson, AZ: SKS Enterprises, 2003.

Scott, Sir James George. *Burma from the Earliest Times to the Present Day (The Story of the Nations).* London: T. Fisher Unwin, 1924.

Tavernier, Jean Baptiste. *Travels in India.* William Crooke, ed. New Delhi: Munshiram Manohavial Publishers, 2001.

Thompson, C.J.S. *The Mysteries and Secrets of Magic.* New York: Causeway Books, 1973.

Webster, R., and P.G. Read. *Gems: Their Sources, Descriptions and Identification.* Fifth edition. Oxford: Butterworth-Heinemann, 1994.

Wise, Richard W. *Secrets of the Gem Trade: The Connoisseur's Guide to Precious Gemstones.* Lenox, MA: Brunswick House Press, 2003.

Yule, H., and H. Cordier. *The Book of Ser Marco Polo.* London: Murray, 1920. Reprint, Dover, 1993.

Index

About the Author

DIANE MORGAN is Adjunct Professor of Religion and Philosophy, Department of Religion and Philosophy, Wilson College. She is the author of over thirty books including *From Satan's Crown to the Holy Grail: Emeralds in Myth, Magic, and History* (Praeger, 2007) and *The Buddhist Experience in America* (Greenwood Press, 2004).